180 DAYS AROUND THE WORLD

Learning About Countries & Cultures Through Research & Thinking-Skills Activities

by Shirley Cook

Incentive Publications, Inc.
Nashville, Tennessee

Cover by Geoffrey Brittingham
Illustrated by Marta Johnson
Edited by Jan Keeling

ISBN 0-86530-253-7

Table of Contents

Teachers' Notes 9

WORLD MAPS

Africa 13
Asia 14
Australia...15
Central America...16
Europe..17
North America 18
South America 19

WORLD JOURNEY

1. Chad 23
2. Mongolia 24
3. Cyprus 25
4. Scotland...................26
5. Virginia...................27
6. Armenia...................28
7. Kenya 29
8. Nova Scotia 31
9. Washington, D.C. 32
10. Cape Verde 33
11. Mexico 34
12. Portugal 36
13. New Mexico 38
14. Bahamas 39
15. Latvia 40
16. Thailand 41
17. Zimbabwe................42
18. Norway...................43
19. Canary Islands.........45
20. Lebanon..................46
21. Rwanda...................47
22. Venezuela 48
23. Albania 50
24. Ivory Coast.............51
25. France 52
26. India 53

27. Senegal 54
28. Wisconsin 55
29. Manitoba 56
30. Chile......................57
31. Gambia 58
32. Bangladesh 59
33. Kuwait....................60
34. Sierra Leone 61
35. California 62
36. Bermuda 63
37. Guatemala 64
38. Montana 65
39. Syria 66
40. Peru 67
41. Maine 68
42. British Columbia........69
43. Guyana 70
44. Libya 71
45. Turkey 72
46. Lesotho...................73
47. Botswana 74
48. Ecuador 75
49. Laos 76
50. Trinidad and Tobago 77
51. Western Sahara..........78
52. Cambodia (Kampuchea) 79

53. China	80	
54. Indonesia	81	
55. Oman	82	
56. Texas	83	
57. Zambia	84	
58. Quebec	85	
59. Guinea-Bissau	86	
60. Louisiana	87	
61. Somalia	88	
62. Yugoslavia	89	
63. Colorado	90	
64. Bosnia	91	
65. Liberia	92	
66. San Marino	94	
67. Vietnam	95	
68. Algeria	96	
69. Guinea	97	
70. Madagascar	98	
71. Puerto Rico	99	
72. Pennsylvania	100	
73. Illinois	101	
74. Alberta	102	
75. Greenland	103	
76. Jamaica	104	
77. Phillipines	105	
78. Florida	106	
79. Wales	107	
80. Ireland	108	
81. American Samoa	109	
82. Costa Rica	110	
83. Malta	112	
84. Pakistan	113	
85. New York	114	
86. Barbados	116	
87. Estonia	118	
88. Korea	119	
89. Sudan	120	
90. Zaire	121	
91. Massachusetts	122	
92. Burma	123	
93. Malaysia	124	
94. Germany	125	
95. Romania	126	
96. Malawi	127	
97. Angola	128	
98. Ghana	129	
99. Iran	130	
100. South Africa	131	
101. Minnesota	133	
102. Brazil	134	
103. Gabon	136	
104. Jordan	137	
105. Mauritius	138	
106. Central African Republic	139	
107. Greece	140	
108. Mauritania	141	
109. Sweden	142	
110. Rhode Island	144	
111. Burkina Faso	145	
112. Egypt	146	
113. Japan	147	
114. Spain	149	
115. Argentina	150	
116. Panama	151	
117. Israel	152	
118. Colombia	153	
119. Brunei	155	
120. Finland	156	
121. Comoros	157	
122. Luxembourg	158	
123. Suriname	159	
124. Cameroon	160	
125. Haiti	162	
126. Morocco	163	
127. Uganda	164	
128. Connecticut	165	
129. Iceland	166	
130. Nicaragua	167	

131. Taiwan 168
132. Arizona 169
133. Cuba 170
134. Italy 172
135. Hawaii 173
136. Ontario 174
137. Honduras 175
138. Netherlands 176
139. Tunisia 177
140. Belgium 178
141. Ethiopia 179
142. Lithuania 180
143. Sri Lanka 181
144. Bulgaria 182
145. Hungary 183
146. New Zealand 184
147. United States of America 185
148. Bolivia 186
149. Delaware....................187
150. Iraq 188
151. Paraguay....................189
152. Benin 190
153. Niger....................191
154. Tonga....................192
155. Australia....................193

156. Denmark 196
157. Mali 197
158. Swaziland 198
159. Austria 199
160. Georgia 200
161. Monaco....................201
162. Tanzania 203
163. Newfoundland 204
164. Idaho....................205
165. Bahrain 206
166. Mozambique....................207
167. Hong Kong 208
168. Utah....................209
169. Dominican Republic....................210
170. Nepal 211
171. Nigeria....................212
172. Fiji 213
173. England 214
174. Saudi Arabia....................216
175. Poland....................217
176. Switzerland....................218
177. Alaska....................219
178. Djibouti 220
179. El Salvador 221
180. Congo 222

CREATIVE THINKING AND RESEARCH PAGES

Research Activity Sheet....................225
Artifact Box 226
Changing Events....................227
What Time Is It? 228
Latitude, Longitude, And Distance....................230
Mathematically Speaking 232
Flight Relay 233
Say What? (Language) 234
Eating Goober Peas 236
Significance Search 237

Answer Keys....................238
Alphabetical List Of Locations....................239

Teachers' Notes

As educators we know that children retain little of what they hear, some of what they see, and much more of what they do. We continually strive to provide all of our students with an opportunity to "learn how to learn" rather than to simply memorize facts. This book of global challenges provides all children with opportunities "to do" rather than simply to listen or watch.

180 Days Around The World gives students a wonderful opportunity to learn about 180 of the world's most interesting countries, provinces, and states and the people who inhabit them, while using genuine research skills. Research is not merely the copying of information from a reference book to the student's paper—it is the quest for an answer, the solving of a mystery, if you will. Further, global knowledge is more than the ability to locate our neighboring states. It is the ability to put together the entire scope of the world's bodies of land in a meaningful way and to relate to the diverse cultures of the world.

The format of this book enables teachers to use it in a variety of ways. One way would be to begin each day by giving the children, singly or in cooperative groups, a Location Card. During the day, the children do some research to find the country, state, or province to which the clues on the card refer. The solution could be discussed later in the day, with time for children to become involved in the other activities connected with each location. Another way to use these materials would be to give out one clue at a time, allowing the children to narrow down the possibilities until the final clue allows for positive identification. Or, post the clue card in a familiar place at the beginning of the day, allowing children to begin their search for the mystery location as soon as they arrive. A child's delight in solving mysteries will motivate him or her to find each location. Once a student becomes aware of the different types of reference materials available, the clues will not only provide information of global significance, but will strengthen the student's ability to learn through inquiry.

After the clues are researched, it is a good time for the teacher to discuss some of the clues with the class and to determine how various members of the group arrived at their answers. Reference works and facts about the country can be highlighted here. Attention to maps and globes will provide further international awareness. Children may choose to keep track of each country that

9

they have found by placing a mark, such as a colored dot, on the maps on pages 13-19. You may provide individual copies of the maps for students, or use an opaque projector to enlarge the maps for display on a wall or bulletin board. At the end of the school year, the children will marvel at the places they have traveled and the people they have learned about.

Children who need additional challenges can follow up their research with other activities provided for each country. These materials are designed to integrate Social Studies with other areas of the curriculum, enhance thinking skills, and provide additional opportunities for research.

This book gives children an opportunity to travel around the world during the course of the school year, to gain further information about important world locations and the people who inhabit them, to enhance research skills, and to remain highly motivated by the learning process. While this is not a complete Social Studies curriculum, it provides students with vital global awareness in a challenging and meaningful way while integrating global awareness and cultural diversity with other areas of the curriculum.

WORLD MAPS

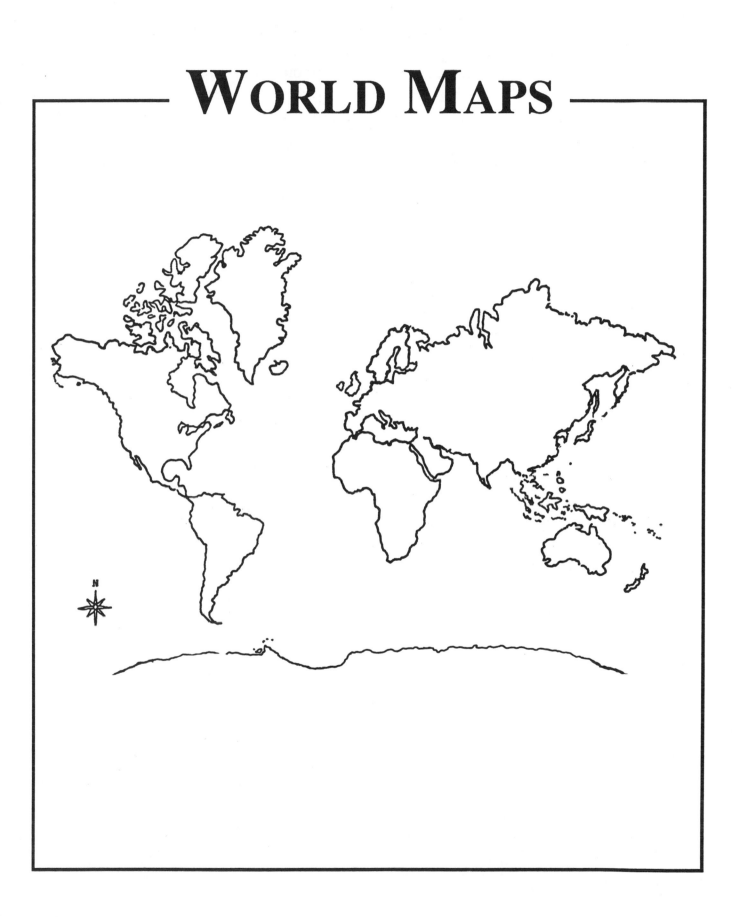

AFRICA

ASIA

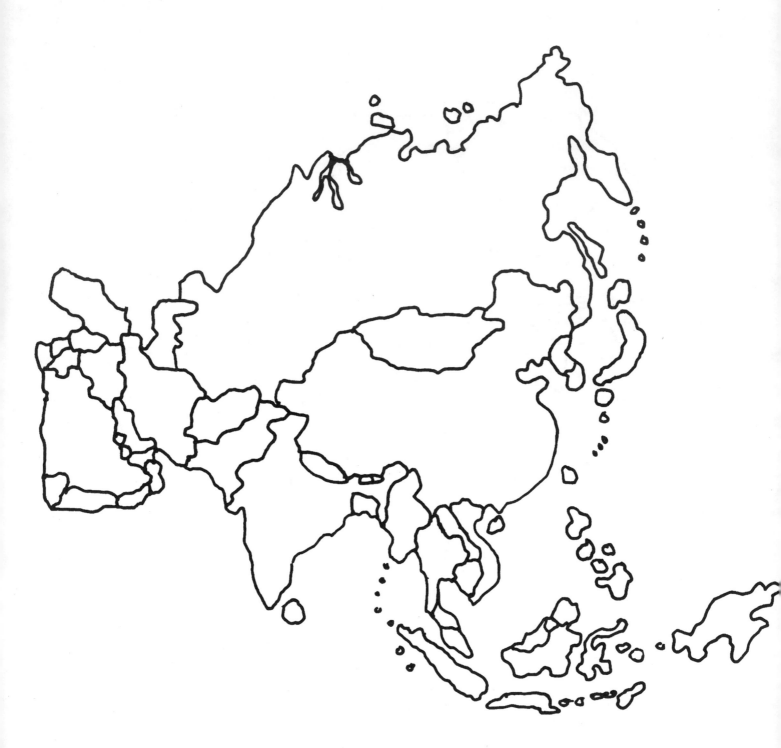

AUSTRALIA

CENTRAL AMERICA

EUROPE

NORTH AMERICA

SOUTH AMERICA

WORLD JOURNEY

Name _____

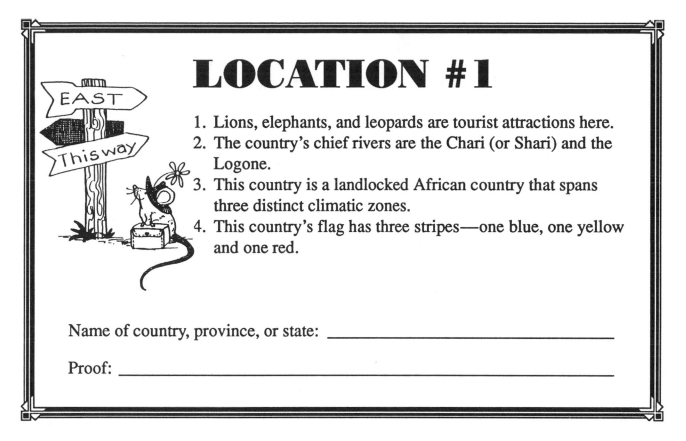

LOCATION #1

1. Lions, elephants, and leopards are tourist attractions here.
2. The country's chief rivers are the Chari (or Shari) and the Logone.
3. This country is a landlocked African country that spans three distinct climatic zones.
4. This country's flag has three stripes—one blue, one yellow and one red.

Name of country, province, or state: _____

Proof: _____

Creativity Across The Curriculum

1. Enlarge the outline of this country on a piece of white paper. Using magazines or newspapers, cut out pictures that represent this country and glue them inside the outline.

2. A mandala is a drawing inside a circle that is basically composed of geometric figures. It focuses attention on color and design rather than on reality. Create your own elaborate mandala and do some research to see if you can discover the origin of this type of art.

Name _____

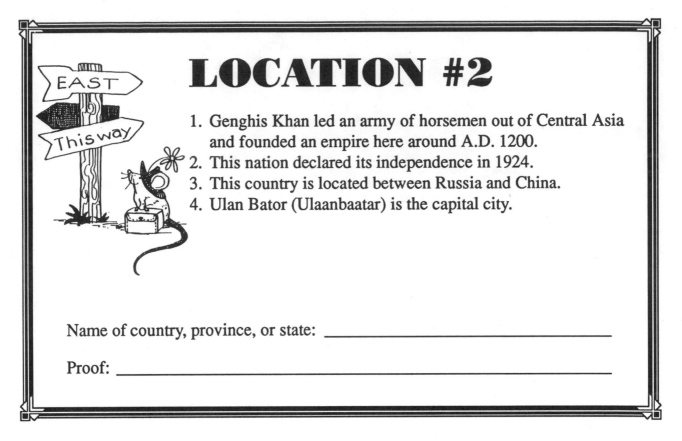

LOCATION #2

1. Genghis Khan led an army of horsemen out of Central Asia and founded an empire here around A.D. 1200.
2. This nation declared its independence in 1924.
3. This country is located between Russia and China.
4. Ulan Bator (Ulaanbaatar) is the capital city.

Name of country, province, or state: _____

Proof: _____

Creativity Across The Curriculum

1. Much of our world has been involved in war at one time or another. Yet we know that people have many qualities of value to society as a whole. Brainstorm a list of good qualities of each of your classmates. Write your list on a separate piece of paper.

2. For hundreds of years Nomads on horseback herded sheep and lived in round tents called "gers" or "yurts." Trucks and motorbikes are now common methods of travel, and many nomads live on state farms and have televisions in their tents. List modern conveniences that these nomads probably still go without. If you could provide them with 5 additional conveniences, what would you choose?

3. We talk about all people being equal.
Are all people equal?
Why or why not?

Name _____

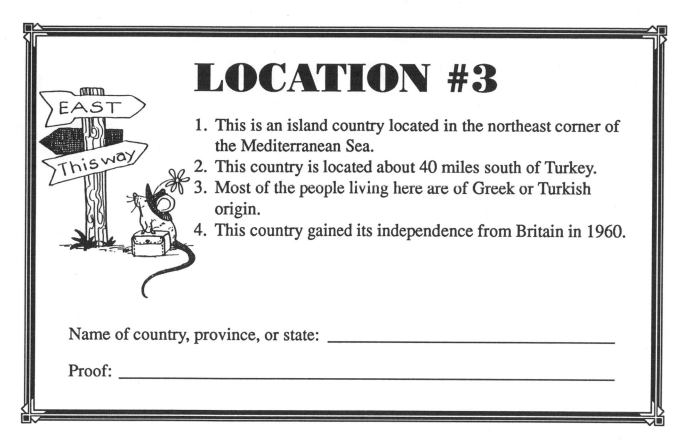

LOCATION #3

1. This is an island country located in the northeast corner of the Mediterranean Sea.
2. This country is located about 40 miles south of Turkey.
3. Most of the people living here are of Greek or Turkish origin.
4. This country gained its independence from Britain in 1960.

Name of country, province, or state: _____

Proof: _____

Creativity Across The Curriculum

1. You and your family are about to sail away to a country of your choice—all expenses paid. Where will you go? There's just one catch—you must remain there for three years. Explain your choice.

2. You have been stranded high atop this island country's mountainous area. Using the dot-dash International Morse Code system, write a message asking for help to send down the mountain by donkey.

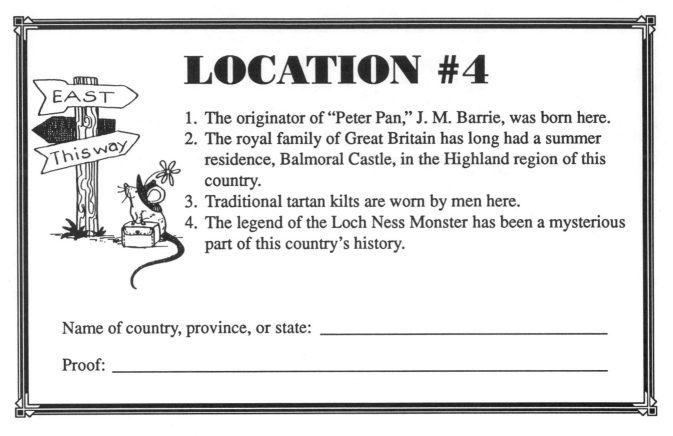

LOCATION #4

1. The originator of "Peter Pan," J. M. Barrie, was born here.
2. The royal family of Great Britain has long had a summer residence, Balmoral Castle, in the Highland region of this country.
3. Traditional tartan kilts are worn by men here.
4. The legend of the Loch Ness Monster has been a mysterious part of this country's history.

Name of country, province, or state: _____

Proof: _____

Creativity Across The Curriculum

1. The game of golf was created in this country in the 12th century. List your ten favorite sports in rank order. If all sports in your country had to be eliminated except one, which one would you choose to keep? Why?

2. Many country folks here have high tea at 5:30 or 6:00. Tea bread is always served, along with bacon and eggs, sausage, or fish and chips. If you were preparing high tea, what would you serve?

3. Mary, Queen of Scots, was beheaded by Queen Elizabeth in 1587. Create a timeline of events that led to Mary's death.

4. Compare "tossing the caber" to a sporting event in the United States.

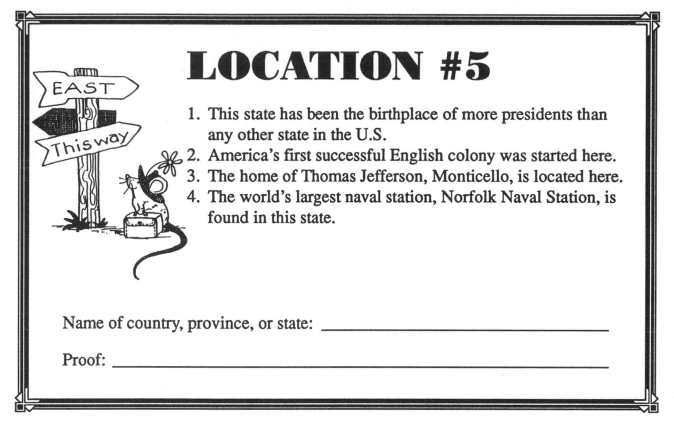

LOCATION #5

1. This state has been the birthplace of more presidents than any other state in the U.S.
2. America's first successful English colony was started here.
3. The home of Thomas Jefferson, Monticello, is located here.
4. The world's largest naval station, Norfolk Naval Station, is found in this state.

Name of country, province, or state: _____

Proof: _____

Creativity Across The Curriculum

1. The college of William and Mary, located in this state, is the second oldest U.S. college. Do some research to find out the name of the oldest college. Tell how it has changed through the years.

2. Tobacco is this state's most valuable crop. It accounts for about two billion dollars in yearly revenue. Write a letter to the governor convincing him that tobacco should no longer be grown in the state.

3. This state has been nicknamed the "Mother of Presidents" because of all of the presidents born here. Do research to identify these presidents. List them in order of importance according to the contributions they have made to our country.

Name _____

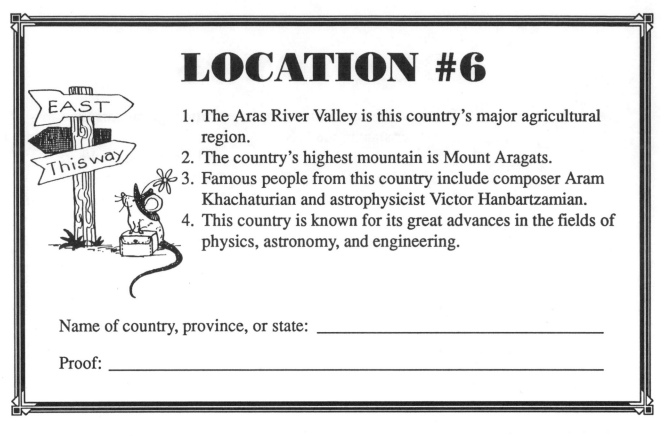

LOCATION #6

1. The Aras River Valley is this country's major agricultural region.
2. The country's highest mountain is Mount Aragats.
3. Famous people from this country include composer Aram Khachaturian and astrophysicist Victor Hanbartzamian.
4. This country is known for its great advances in the fields of physics, astronomy, and engineering.

Name of country, province, or state: _____

Proof: _____

Creativity Across The Curriculum

1. Rice is one of the major crops here. If you had only rice, water, sugar, and spices, what kind of meal would you make for your family?

2. Create a comic strip that will give information about your country by cutting out comic strip characters, arranging them in a meaningful sequence, and giving them captions.

3. Do some research to find out the type of work that an astrologist has to do.

4. Create a list of five questions about this country. Ask a friend to find the answers.

Name _____

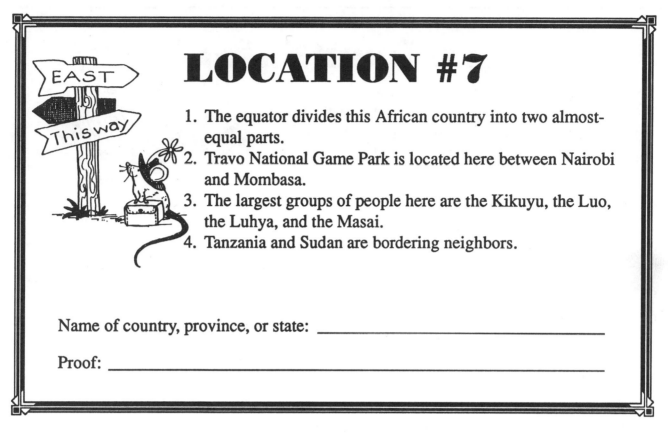

LOCATION #7

1. The equator divides this African country into two almost-equal parts.
2. Travo National Game Park is located here between Nairobi and Mombasa.
3. The largest groups of people here are the Kikuyu, the Luo, the Luhya, and the Masai.
4. Tanzania and Sudan are bordering neighbors.

Name of country, province, or state: _____

Proof: _____

Creativity Across The Curriculum

1. Cassava, Sisal, and Pyrethrum are some of this country's chief crops. Interview several people and ask them to describe Cassava, Sisal, and Pyrethrum. Prepare a poster to convince people to buy one of these products.

2. Do some research to find out about the Masai way of life.

3. Write an African song.

African Safari

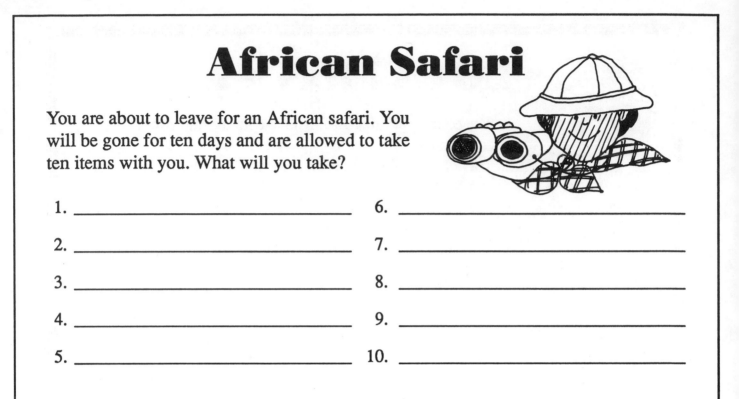

You are about to leave for an African safari. You will be gone for ten days and are allowed to take ten items with you. What will you take?

1. _____   6. _____

2. _____   7. _____

3. _____   8. _____

4. _____   9. _____

5. _____   10. _____

Create a list of animals that live in the jungle. Draw your favorite.

_____

_____

_____

_____

_____

_____

_____

_____

_____

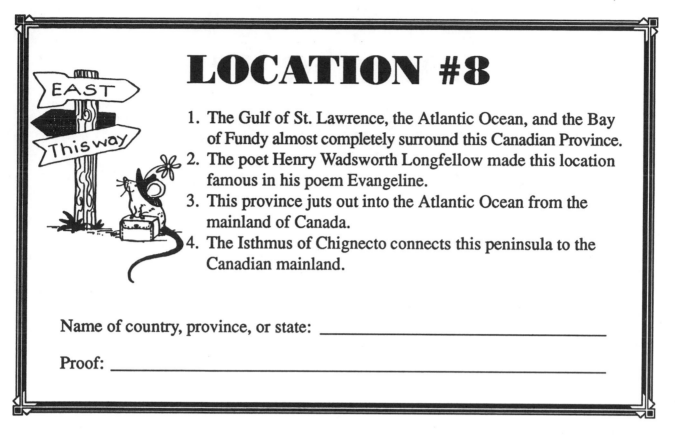

LOCATION #8

1. The Gulf of St. Lawrence, the Atlantic Ocean, and the Bay of Fundy almost completely surround this Canadian Province.
2. The poet Henry Wadsworth Longfellow made this location famous in his poem Evangeline.
3. This province juts out into the Atlantic Ocean from the mainland of Canada.
4. The Isthmus of Chignecto connects this peninsula to the Canadian mainland.

Name of country, province, or state: _____

Proof: _____

Creativity Across The Curriculum

• Longfellow was so inspired by this location that he created a beautiful poem about it. Create a poem of ten lines or longer about the area in which you live.

_____

_____

_____

_____

_____

_____

_____

_____

_____

_____

_____

_____

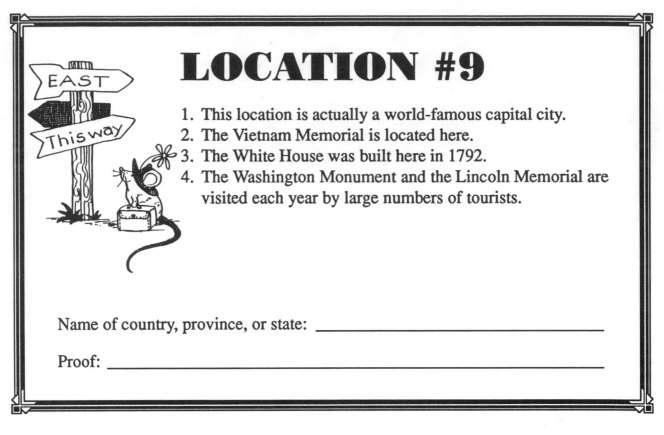

LOCATION #9

1. This location is actually a world-famous capital city.
2. The Vietnam Memorial is located here.
3. The White House was built here in 1792.
4. The Washington Monument and the Lincoln Memorial are visited each year by large numbers of tourists.

Name of country, province, or state: _____

Proof: _____

Creativity Across The Curriculum

1. Presidents Lincoln, Washington, Jefferson, and Kennedy are a few of the presidents honored here through statues, monuments, and grave markers. Choose a favorite president and design an appropriate monument dedicated to his honor. Sketch your monument on another sheet of paper and detail the reasons for your design.

2. No woman has ever been elected to the presidency of the United States. Throughout history, however, many women have made important contributions to our country. On a sheet of paper, list ten women from the past and ten women from the present who you feel would have made good presidential candidates. List your choices in rank order.

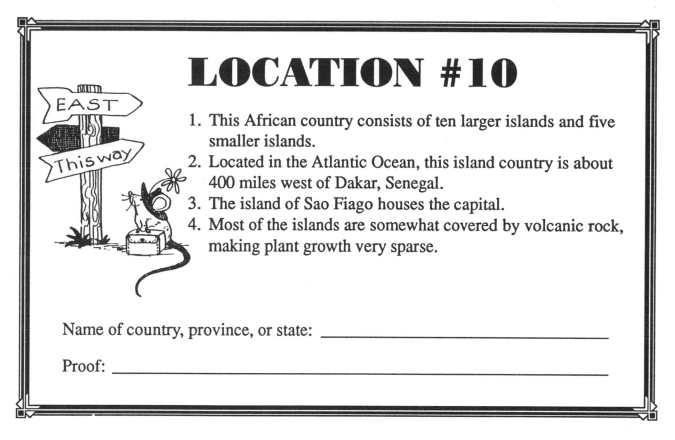

LOCATION #10

1. This African country consists of ten larger islands and five smaller islands.
2. Located in the Atlantic Ocean, this island country is about 400 miles west of Dakar, Senegal.
3. The island of Sao Fiago houses the capital.
4. Most of the islands are somewhat covered by volcanic rock, making plant growth very sparse.

Name of country, province, or state: _____

Proof: _____

Creativity Across The Curriculum

1. The colors found on this country's flag are primarily yellow, red, and green. Using only yellow, red, and blue, create as many colors as you can. Do research to create your own color wheel.

2. This country mines salt and pozzuolana. Do research to find out what pozzuolana is and how it is used.

3. These islands are completely surrounded by ocean waters. Create a picture of the ocean using watercolor paints. What colors will you use for the water?

LOCATION #11

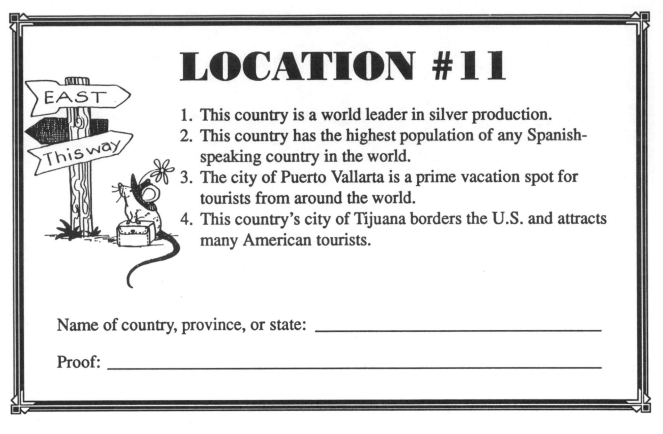

1. This country is a world leader in silver production.
2. This country has the highest population of any Spanish-speaking country in the world.
3. The city of Puerto Vallarta is a prime vacation spot for tourists from around the world.
4. This country's city of Tijuana borders the U.S. and attracts many American tourists.

Name of country, province, or state: _____

Proof: _____

Creativity Across The Curriculum

Create a paper-maché mini-piñata by following these directions:

1. Inflate several small balloons. Tape them into the desired shape. Put crumpled newspaper in any gaps.
2. Cut newspaper strips 1" wide. Dip in flour-and-water paste. Cover the form with layers of newspaper strips. Let this dry for 48 hours.
3. Add more strips and again dry thoroughly. Cut a small door in the bottom. Remove the balloons.
4. Decorate the pinata by adding small strips of tissue paper. (Construction and other types of paper also work.)
5. Fill with goodies and hinge the door shut with tape.
6. While blindfolded, friends may use a plastic bat to try to open the piñata.

• The sombrero is the hat associated with this country. On another piece of paper, sketch and color five other types of hats that are closely linked to specific countries.

• List ten new ways to use a tortilla, a piñata, and a sombrero.

1. _____   6. _____

2. _____   7. _____

3. _____   8. _____

4. _____   9. _____

5. _____   10. _____

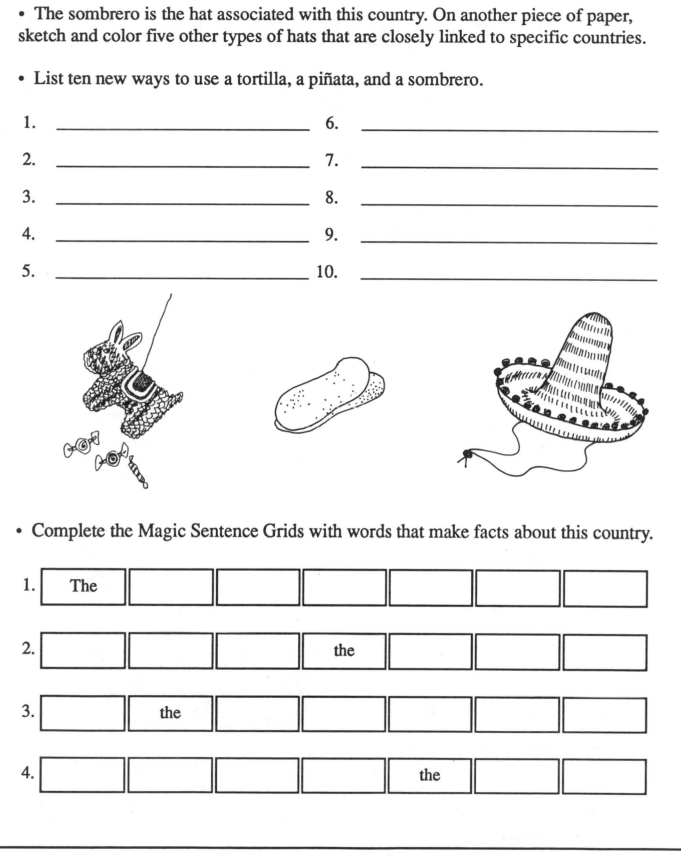

• Complete the Magic Sentence Grids with words that make facts about this country.

1.	The						

2.				the			

3.		the					

4.					the		

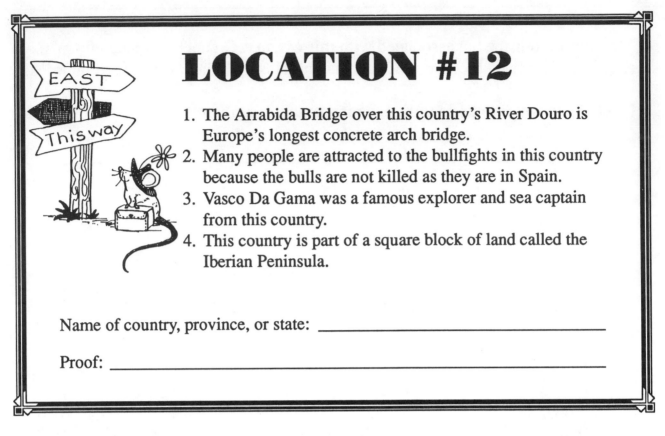

LOCATION #12

1. The Arrabida Bridge over this country's River Douro is Europe's longest concrete arch bridge.
2. Many people are attracted to the bullfights in this country because the bulls are not killed as they are in Spain.
3. Vasco Da Gama was a famous explorer and sea captain from this country.
4. This country is part of a square block of land called the Iberian Peninsula.

Name of country, province, or state: _____

Proof: _____

Creativity Across The Curriculum

1. To land in the capital of this mystery country, complete the activity "Where in the World Am I?" (page 37).

2. Choose a topic that you enjoy. Create a list of words that you think are associated with your topic. Then create a riddle that would allow others to guess your topic.

3. Put together a five-animal food chain that would be found in this country or in the ocean nearby.

Where In The World Am I?

Begin in Pretoria, South Africa. Travel due north for approximately 550 miles. The

city you come to is _____. It is located in the country

of _____. Again, travel due north approximately 1070

miles, then proceed west approximately 1400 miles. Progress along the coastline in a

north to northwest direction until you reach the next capital city. This country is

called _____. Continue east about 1560 miles. Your

next stop is the city of _____. Now proceed 1070 miles north

to the city of _____. The number of miles you will travel to

reach Tripoli is _____. Continue traveling west to the country

approximately 1000 miles away. This country is _____.

Proceed north to the nearest capital. This capital is named _____.

Finally, cross a body of water to the nearest northwest capital city. You have landed

in _____.

LOCATION #13

EAST
This way

1. The United States' oldest public building, the Palace of the Governors, is located here.
2. The world's first atomic bomb was set off here in 1945.
3. Carlsbad Caverns National Park is found here.
4. Many pueblos, Indian cliff houses, are found here.

Name of country, province, or state: _____

Proof: _____

Creativity Across The Curriculum

1. The desert area of this state was used for Robert Goddard's rocket experiments. Write an editorial to convince people of the value of spending billions of dollars on the space program even though we live in an age of homelessness, high taxes, and diseases without cures—or, write an editorial to convince people that money should *not* be spent on the space program.

2. Pueblo Bonito, located here, has 800 rooms. If you lived in a house with 800 rooms, how would you use those rooms? How many bedrooms, bathrooms, etc., would you have?

LOCATION #14

EAST
This way

1. This country is a chain of about 3,000 islands and reefs located in the West Indies.
2. Approximately 20 of the islands are inhabited.
3. As these islands are only fifty miles off the coast of Florida, their cities, such as Nassau, are often visited by American cruise ships.
4. This country gained its independence from Britain on July 10, 1973.

Name of country, province, or state: _____

Proof: _____

Creativity Across The Curriculum

1. You have been given a small island of your own. It is approximately 55 miles off the Florida coast. Bananas, pineapples, and citrus fruits grow here, and fish abound in the surrounding ocean waters. What will you do with your island?

2. How many ways can you prepare a banana?

3. If you lived on an island away from the mainstream of civilization, a first-aid kit would be very important to you. Describe the contents of your first-aid kit and support your choices.

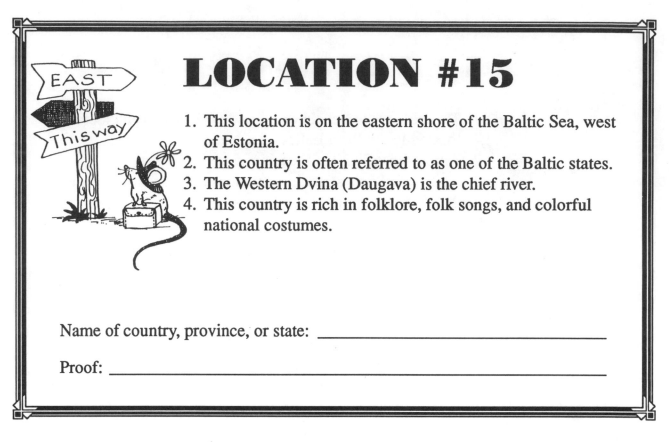

LOCATION #15

1. This location is on the eastern shore of the Baltic Sea, west of Estonia.
2. This country is often referred to as one of the Baltic states.
3. The Western Dvina (Daugava) is the chief river.
4. This country is rich in folklore, folk songs, and colorful national costumes.

Name of country, province, or state: _____

Proof: _____

Creativity Across The Curriculum

1. When the Soviets took control of this country, they demanded many cultural changes—changes in language, religion, news reporting, television programming, jobs, and fashion. If another country took over political leadership of the United States, what five practices associated with your family's cultural heritage would you most wish to keep?

Example: 1. Easter egg hunts

2. The chief industries here produce electronic equipment, household appliances, and machinery. These industries are all changing rapidly. Choose one and create a time line showing how products have changed over the years.

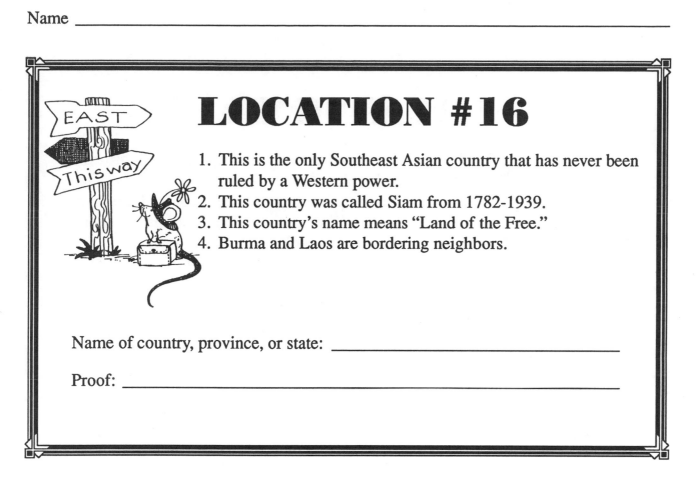

LOCATION #16

1. This is the only Southeast Asian country that has never been ruled by a Western power.
2. This country was called Siam from 1782-1939.
3. This country's name means "Land of the Free."
4. Burma and Laos are bordering neighbors.

Name of country, province, or state: _____

Proof: _____

Creativity Across The Curriculum

1. This nation's Buddhist temples exemplify its architecture. Does the architecture of the major churches of the United States vary greatly from one religious group to another? For example, do Catholic churches generally look much different from Jewish Synagogues? Find pictures or make sketches to compare the architecture of churches of the following religious groups: 1. Catholic 2. Lutheran 3. Jewish 4. Baptist 5. Mormon.

2. Three popular sports or games here are: Thai-style boxing, takraw, and mak ruk. Explain how one of these sports is played. If you could participate in one, which would you choose and why?

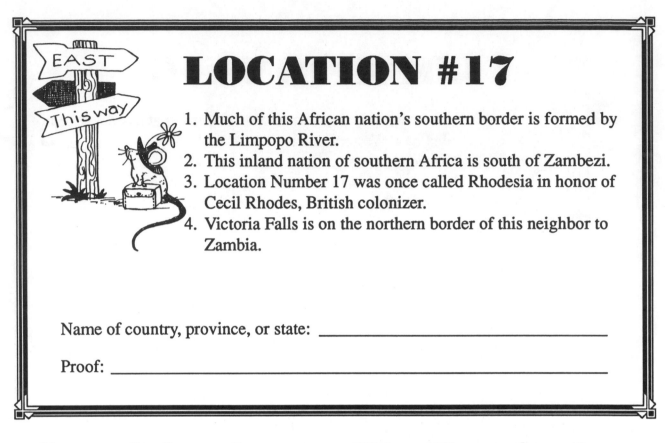

LOCATION #17

1. Much of this African nation's southern border is formed by the Limpopo River.
2. This inland nation of southern Africa is south of Zambezi.
3. Location Number 17 was once called Rhodesia in honor of Cecil Rhodes, British colonizer.
4. Victoria Falls is on the northern border of this neighbor to Zambia.

Name of country, province, or state: _____

Proof: _____

Creativity Across The Curriculum

1. The Acropolis, a hilltop fort built by the Bantu centuries ago, still stands in part today. Compare this hilltop system of defense to our system of defense in the U.S.

2. Many African nations have large numbers of poor people. Should the people of the U.S. feel responsible for helping to feed these undernourished African people? Defend your decision by writing a Letter to the Editor.

LOCATION #18

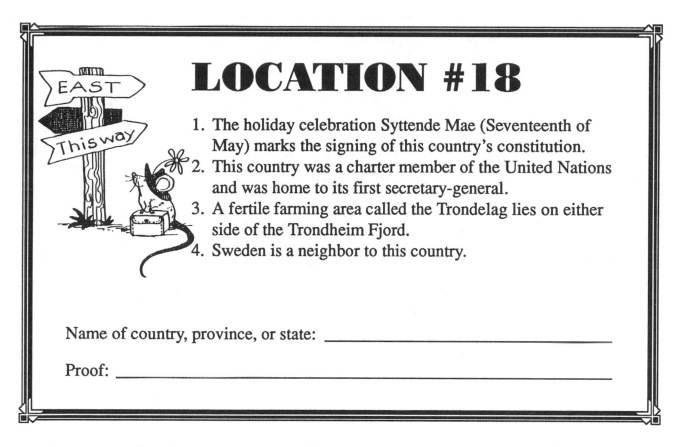

1. The holiday celebration Syttende Mae (Seventeenth of May) marks the signing of this country's constitution.
2. This country was a charter member of the United Nations and was home to its first secretary-general.
3. A fertile farming area called the Trondelag lies on either side of the Trondheim Fjord.
4. Sweden is a neighbor to this country.

Name of country, province, or state: _____

Proof: _____

Creativity Across The Curriculum

1. The first man to reach the South Pole, Roald Amundsen, came from this country. Create a persuasive speech to convince your classmates that this was an important feat.

2. Archaeologists have learned quite a bit about the Vikings. Create a factual (non-fiction) book about the Vikings' discovery of a new land.

3. Rosemaling is one of this country's best known crafts. Using tempera or acrylic paint and some research, try your hand at a flower and leaf in a rosemaling technique.

• This country is famous for its troll tales, folktales, and sagas. A familiar troll tale is "The Three Billy Goats Gruff." Find out about a folktale character such as Tan-Verk-Trollet, the tiny toothache troll, and tell the story to your class as a guest storyteller.

• During the fourteenth century, a disease known as the Black Plague killed hundreds of thousands of people in many European nations. What other diseases have caused tremendous loss of human life throughout history? Rank the top ten from most to least destructive.

1. _____

2. _____

3. _____

4. _____

5. _____

6. _____

7. _____

8. _____

9. _____

10. _____

Name _____

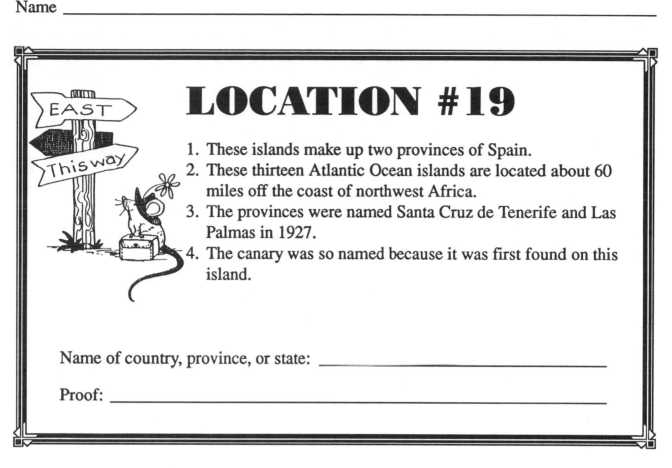

LOCATION #19

1. These islands make up two provinces of Spain.
2. These thirteen Atlantic Ocean islands are located about 60 miles off the coast of northwest Africa.
3. The provinces were named Santa Cruz de Tenerife and Las Palmas in 1927.
4. The canary was so named because it was first found on this island.

Name of country, province, or state: _____

Proof: _____

Creativity Across The Curriculum

1. Make a class survey to determine the class favorite pet. Graph your results. Then carefully plan to advertise the canary as a "perfect pet" for one week. Create signs, banners, and other types of ads. At the end of the week-long campaign, survey your audience again. How do these results compare with those of the first survey?

2. The people living on the island of Gomera communicate with each other over distances with a whistled language somewhat like Spanish. Can you create a sentence in whistled English?

3. In the United States, we fear crime as one of our nation's biggest problems. Do you think crime would be as big a problem on a small island country? Why or why not?

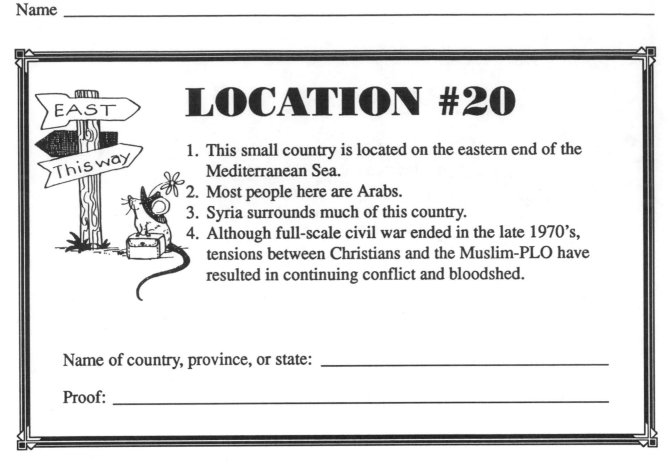

LOCATION #20

1. This small country is located on the eastern end of the Mediterranean Sea.
2. Most people here are Arabs.
3. Syria surrounds much of this country.
4. Although full-scale civil war ended in the late 1970's, tensions between Christians and the Muslim-PLO have resulted in continuing conflict and bloodshed.

Name of country, province, or state: _____

Proof: _____

Creativity Across The Curriculum

1. Different groups in this country have taken foreigners as hostages. If a friend or family member of yours has been wishing to travel to Location Number 20, how can you persuade them not to go?

2. Life in a war-torn country can be sad and depressing. If you lived in a country continually facing danger, what things would you do each day to remain cheerful and optimistic? List ten things.

3. Design a peace banner. Create a slogan that will be catchy and that will encourage all nations to think about the need for people to live together in harmony.

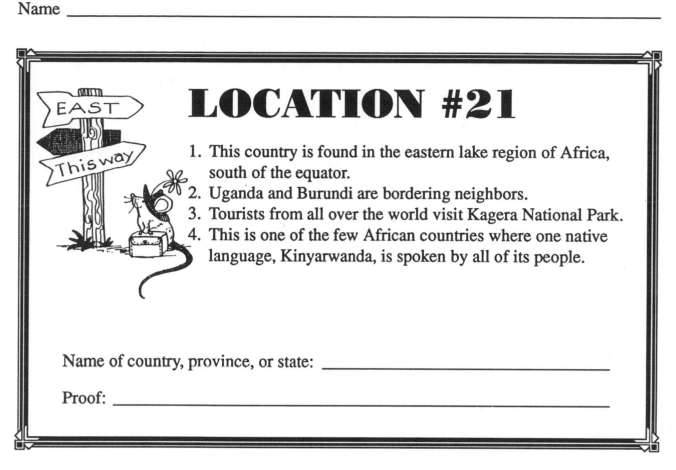

LOCATION #21

1. This country is found in the eastern lake region of Africa, south of the equator.
2. Uganda and Burundi are bordering neighbors.
3. Tourists from all over the world visit Kagera National Park.
4. This is one of the few African countries where one native language, Kinyarwanda, is spoken by all of its people.

Name of country, province, or state: _____

Proof: _____

Creativity Across The Curriculum

1. Bush cats are found in Kagera National Park. Do some research to find out more about these animals. Then design a poster that could be used to advertise their arrival at the local zoo.

2. The first people to live in this country were the Twa, pygmies who grow no taller than five feet. List disadvantages of living in a society where the tallest person is less than five feet tall. What are the advantages?

Name _____

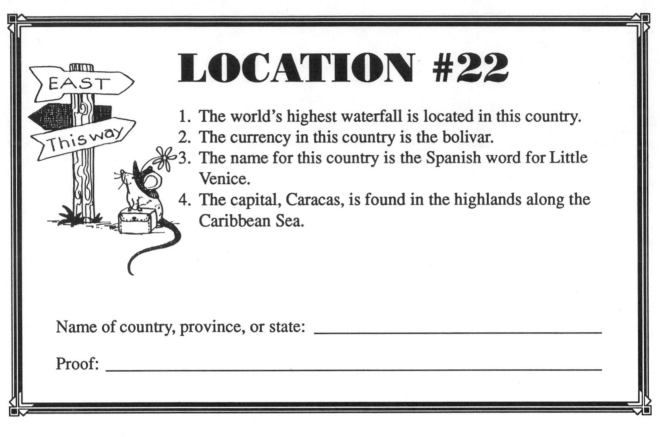

LOCATION #22

1. The world's highest waterfall is located in this country.
2. The currency in this country is the bolivar.
3. The name for this country is the Spanish word for Little Venice.
4. The capital, Caracas, is found in the highlands along the Caribbean Sea.

Name of country, province, or state: _____

Proof: _____

Creativity Across The Curriculum

1. What types of fish would most likely be found in the waters of the Caribbean Sea? Do research to find important facts about two of these types, and sketch detailed pictures.

2. New Orleans has its Mardi Gras and Pasadena has its Rose Parade. Create a country-wide celebration for Country Number 22. What would be its special features? When and why would it be celebrated? Why would people want to return to the festival year after year?

3. Sketch the basic shape of this country onto a 9 x 11 piece of white construction paper. Create something entirely new from the shape. Color it.

Using pictures and/or words, compare and contrast North American and South American cowboys past and present.

	North America	South America
Past		
Present		

• Indians once lived off the coast of this country in houses built on stilts. What would be the advantages of living in a house on the water? How would your life be different if you lived in a stilt house?

LOCATION #23

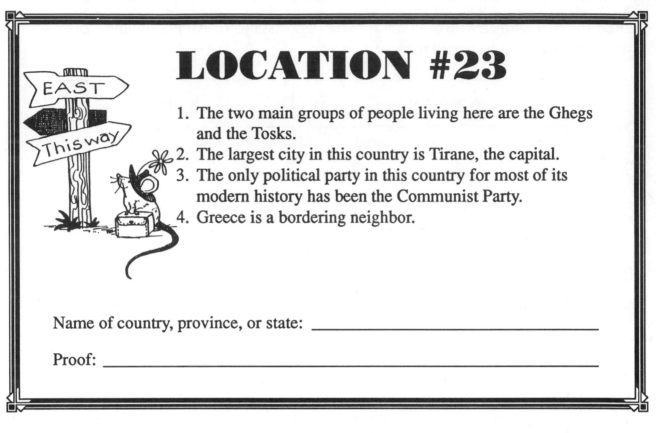

1. The two main groups of people living here are the Ghegs and the Tosks.
2. The largest city in this country is Tirane, the capital.
3. The only political party in this country for most of its modern history has been the Communist Party.
4. Greece is a bordering neighbor.

Name of country, province, or state: _____

Proof: _____

Creativity Across The Curriculum

1. One of the fashions of yesterday that is still worn by men in this country is the Turkish fez. Redesign the fez to update it for today's world.

2. Create five math problems of three-digit numbers that involve mileage in and around Country Number 23.

3. This is a very mountainous country. Create a small salt map to detail the terrain.

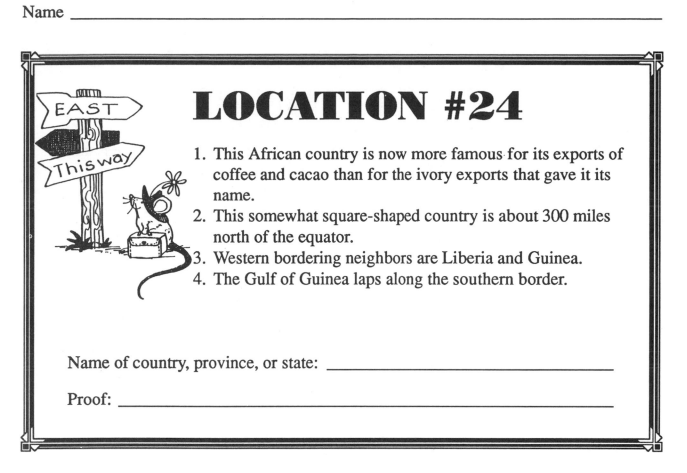

LOCATION #24

1. This African country is now more famous for its exports of coffee and cacao than for the ivory exports that gave it its name.
2. This somewhat square-shaped country is about 300 miles north of the equator.
3. Western bordering neighbors are Liberia and Guinea.
4. The Gulf of Guinea laps along the southern border.

Name of country, province, or state: _____

Proof: _____

Creativity Across The Curriculum

The savannah of the southeast and central parts of this country supports a wide variety of plant and animal life. Sketch each of the following types of birds and frame each with a fact.

1. kori bustard
2. oxpecker
3. ostrich
4. bulbul
5. shrikes
6. cranes

Name _____

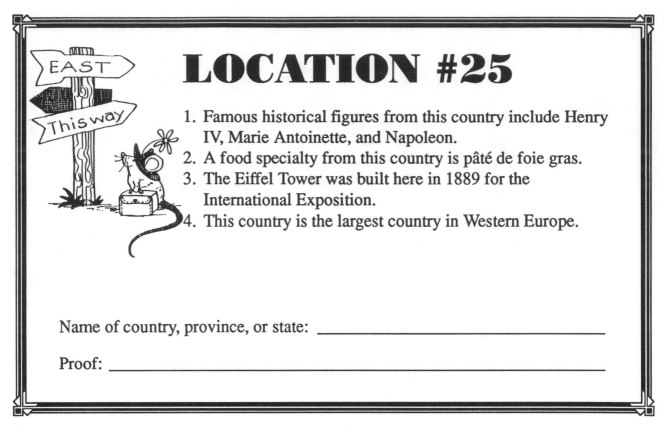

LOCATION #25

1. Famous historical figures from this country include Henry IV, Marie Antoinette, and Napoleon.
2. A food specialty from this country is pâté de foie gras.
3. The Eiffel Tower was built here in 1889 for the International Exposition.
4. This country is the largest country in Western Europe.

Name of country, province, or state: _____

Proof: _____

Creativity Across The Curriculum

1. The Arc de Triomphe is the largest arch in the world, built to honor Napoleon's victories. Build a large and sturdy arch of toothpicks. Measure the height and width of your structure.

2. Petanque is a bowling game played with iron balls that are thrown into the air. Create another game of bowling and write out its instructions.

3. This country is noted for its high fashion. Draw a new fashion design for a boy or girl your age. Then create an advertisement for it. Be sure to include the price.

Name _____

LOCATION #26

1. The people of this country do not eat beef, as the cow is considered a holy animal.
2. This country's famous Taj Mahal was built outside the city of Agra by Shah Jahan, in memory of his wife.
3. For much of its history, a "caste system" divided these people into groups of wealthy and poor.
4. Bangladesh is an eastern neighbor.

Name of country, province, or state: _____

Proof: _____

Creativity Across The Curriculum

1. In parts of this country, fathers may bid for husbands for their daughters by offering a dowry—money and goods—to the prospective husband. If the father wants his daughter to marry an educated man, it will cost thousands of rupees. How would life in the United States change if this system of dowries for husbands existed here? Brainstorm a list of changes that would take place.

2. If you lived in this country, you would not eat beef. Create a new type of fast-food restaurant that would use the major foods found here. What would you call your restaurant, and what would you serve? Design a business logo.

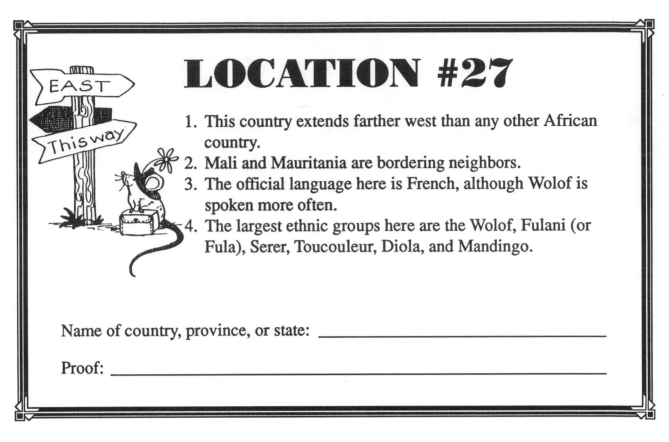

LOCATION #27

1. This country extends farther west than any other African country.
2. Mali and Mauritania are bordering neighbors.
3. The official language here is French, although Wolof is spoken more often.
4. The largest ethnic groups here are the Wolof, Fulani (or Fula), Serer, Toucouleur, Diola, and Mandingo.

Name of country, province, or state: _____

Proof: _____

Creativity Across The Curriculum

1. Kora are stringed instruments made of gourds. They are decorated with brightly colored cloths and are played for pleasure as well as on special occasions such as the Independence Day Festival.

2. The people of this country are known for the striking wooden carved masks they make. Create a mask design on a 12" X 18" piece of paper that you feel could be displayed at an art show. The government here often sponsors art exhibits.

3. The biggest cash crop for farmers here is the peanut. List 25 uses for peanuts. Let your imagination go wild!

LOCATION #28

1. The first Ringling Brothers show was given in this U.S. state in 1882.
2. The first statewide bicycle trail was developed in this midwestern location.
3. This state produces more cheese and butter than does any other U.S. state.
4. Bordering neighbors include Minnesota and Illinois.

Name of country, province, or state: _____

Proof: _____

Creativity Across The Curriculum

1. Do some research to find out how cheese is made. Draw a pictorial representation of the stages in this process.

2. You are about to be interviewed for a job in this state's cheese and butter industry. Tell your prospective boss about four qualities that make you a good candidate for the position and how these qualities will be of value to the company.

3. Hunting and fishing are two popular sports in this state. Write an editorial defending or rejecting the right to hunt animals.

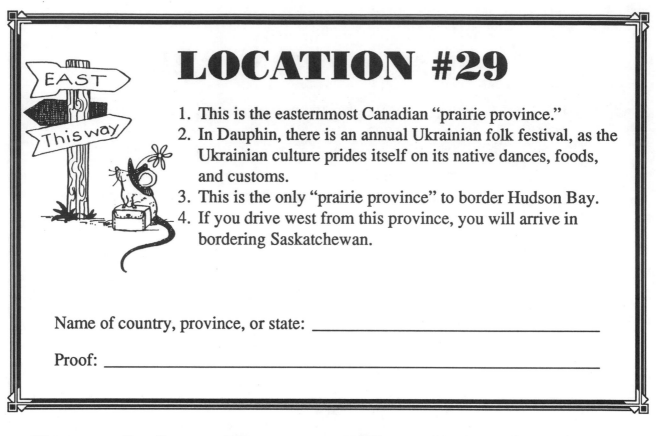

LOCATION #29

1. This is the easternmost Canadian "prairie province."
2. In Dauphin, there is an annual Ukrainian folk festival, as the Ukrainian culture prides itself on its native dances, foods, and customs.
3. This is the only "prairie province" to border Hudson Bay.
4. If you drive west from this province, you will arrive in bordering Saskatchewan.

Name of country, province, or state: _____

Proof: _____

Creativity Across The Curriculum

1. Compare this province to Quebec. List ten differences.

2. When people think of Canada, they think of a peaceful nation. Do some research to find out which wars Canada has been involved in during the past 200 years. Create a time line of your findings.

3. Ukrainian Easter eggs are an important part of this province's cultural heritage. Do research to find out how these eggs look. Then, using a simple egg outline on white paper, create a design that you think would be suitable for a Ukrainian egg.

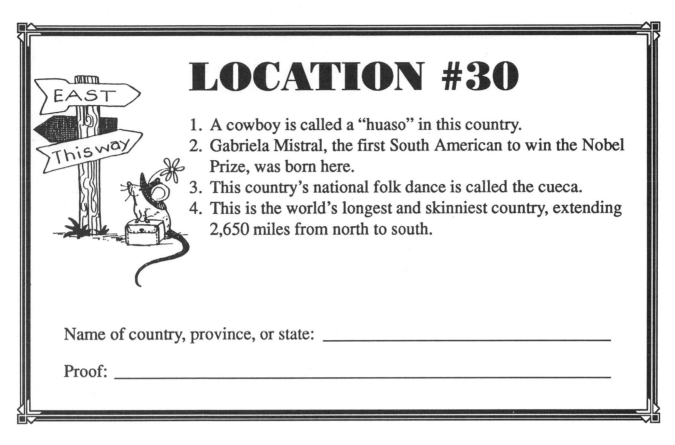

LOCATION #30

1. A cowboy is called a "huaso" in this country.
2. Gabriela Mistral, the first South American to win the Nobel Prize, was born here.
3. This country's national folk dance is called the cueca.
4. This is the world's longest and skinniest country, extending 2,650 miles from north to south.

Name of country, province, or state: _____

Proof: _____

Creativity Across The Curriculum

1. Rayuela is a popular coin pitch game here, although it originated in Spain. Players toss coins or discs toward a line on the ground. The person whose coin lands closest to the line wins. Create a new coin toss game and try it out with a friend.

2. Football is called futbol here. If you were going to spell words the way they sound, how would you spell the following?

    1. height   4. knight
    2. comb   5. pneumonia
    3. gnome   6. telephone

3. *Robinson Crusoe* was written about a man shipwrecked on an island close to this country. Read a portion of *Robinson Crusoe*, by Daniel Defoe.

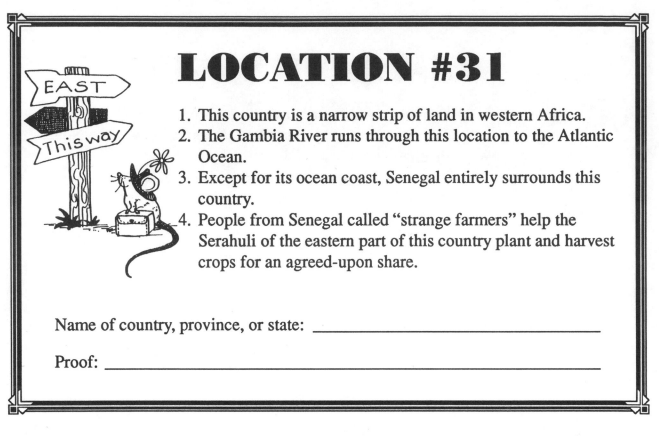

LOCATION #31

1. This country is a narrow strip of land in western Africa.
2. The Gambia River runs through this location to the Atlantic Ocean.
3. Except for its ocean coast, Senegal entirely surrounds this country.
4. People from Senegal called "strange farmers" help the Serahuli of the eastern part of this country plant and harvest crops for an agreed-upon share.

Name of country, province, or state: _____

Proof: _____

Creativity Across The Curriculum

1. Prepare a chart comparing three African cultures. Include language, religion, location, foods, customs, and occupations.

2. Compare the ability to make a living here in the past and the present to making a living past and present in your state.

3. Create a tongue-twister that conveys an important fact about this country.

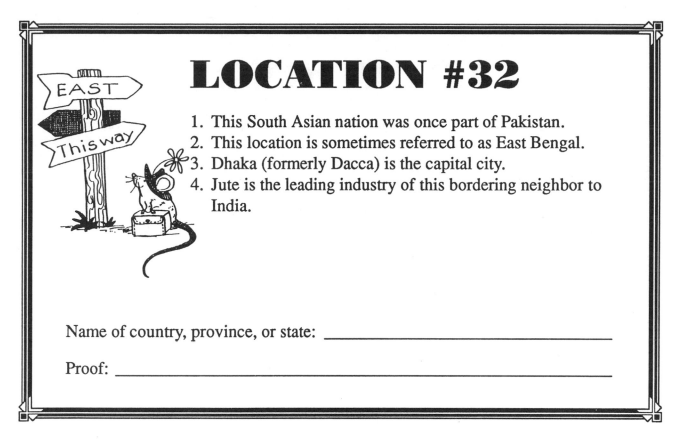

LOCATION #32

1. This South Asian nation was once part of Pakistan.
2. This location is sometimes referred to as East Bengal.
3. Dhaka (formerly Dacca) is the capital city.
4. Jute is the leading industry of this bordering neighbor to India.

Name of country, province, or state: _____

Proof: _____

Creativity Across The Curriculum

1. Cyclones and tidal waves have caused much death and destruction in this country. Choose three types of storms to compare, such as hailstorms, tornadoes, hurricanes, cyclones, and blizzards. Tell about their causes, where they are most likely to occur, the types of destruction possible, when they usually happen, and a brief summary of one of the more famous storms.

2. Research to find out how jute is made. Create a jute "friendship heart" to give to somebody special by following these simple directions:
    1. Cut six pieces of jute that are five feet long.
    2. Divide the jute into two groups of three.
    3. Braid each set of three jute strings three-fourths of the way.
    4. Attach the braids to one another with a rubberband.
    5. Form into a heart shape. Unravel the bottom of the jute. Join the bottom together with another piece of jute. Add a decorative ribbon if desired.

LOCATION #33

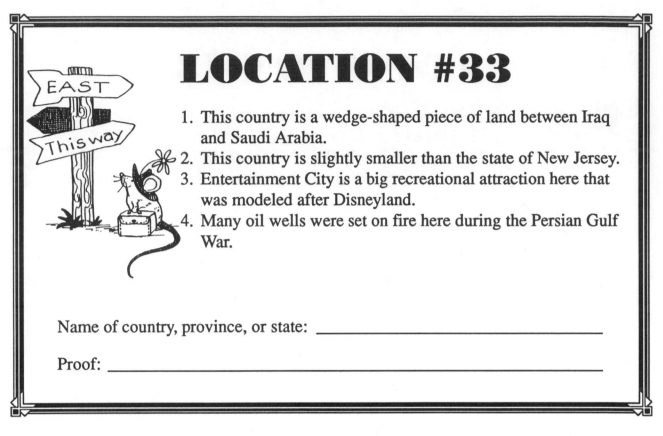

1. This country is a wedge-shaped piece of land between Iraq and Saudi Arabia.
2. This country is slightly smaller than the state of New Jersey.
3. Entertainment City is a big recreational attraction here that was modeled after Disneyland.
4. Many oil wells were set on fire here during the Persian Gulf War.

Name of country, province, or state: _____

Proof: _____

Creativity Across The Curriculum

1. Summer temperatures in this country may exceed 120 degrees Fahrenheit. Design a type of clothing for men or women that would be comfortable in this heat.

2. Ornate gold jewelry is a sign of wealth here. Design a gold hairpiece or large necklace that is very detailed.

3. Pearling still continues as a small-scale industry in this country. Create a diagram to show step-by-step how pearls are extracted from the sea and made into jewelry.

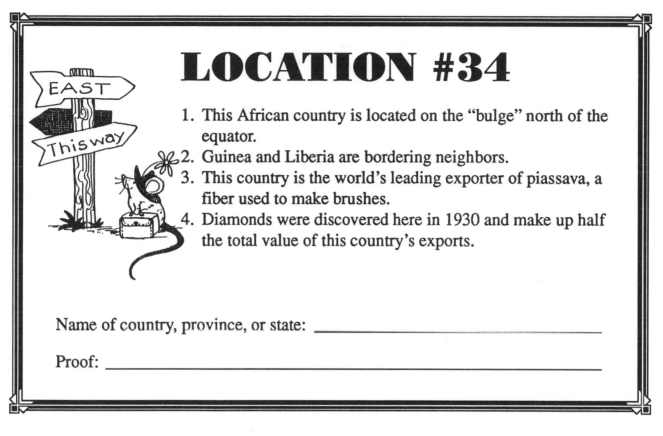

LOCATION #34

1. This African country is located on the "bulge" north of the equator.
2. Guinea and Liberia are bordering neighbors.
3. This country is the world's leading exporter of piassava, a fiber used to make brushes.
4. Diamonds were discovered here in 1930 and make up half the total value of this country's exports.

Name of country, province, or state: _____

Proof: _____

Creativity Across The Curriculum

1. Diamonds are an important export for this country. Do some research to find out how diamonds are priced. What determines the worth of a diamond? Create a diamond necklace designed to scale. Describe it and price it.

2. The law doesn't require the children here to attend school. What effect do you think this has on the economy of the country?

3. A variety of religions are practiced here. The Mende believe the god Ngewo created the world and its contents, such as the valuable nomoli. The nomoli are small human images found in the soil. Mende believe finding a nomali will bring them good crops. Read a book about superstitions. Pick five superstitions and find out how they originated.

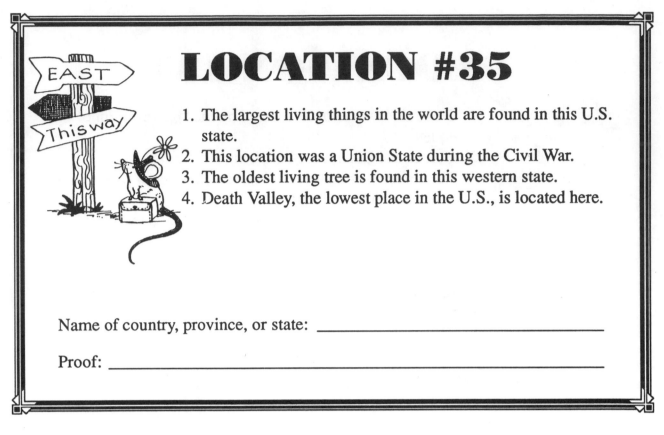

LOCATION #35

1. The largest living things in the world are found in this U.S. state.
2. This location was a Union State during the Civil War.
3. The oldest living tree is found in this western state.
4. Death Valley, the lowest place in the U.S., is located here.

Name of country, province, or state: _____

Proof: _____

Creativity Across The Curriculum

1. List ten important uses for gold. Each must be distinctly different. (For example—rings, bracelets, necklaces, etc., will all count as jewelry.)

2. The condor named after this state is an endangered species. Find out why. What can we do to prevent its extinction? Sketch the bird in an appropriate habitat.

3. Using population figures from major cities located here, create five two-part story problems that involve multiplication, addition, subtraction and/or division.

LOCATION #36

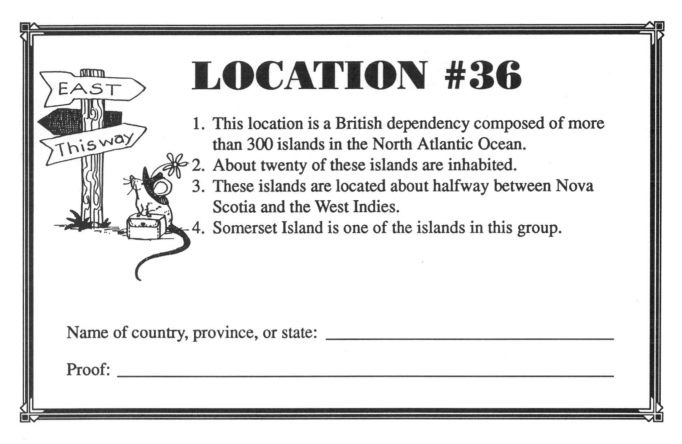

1. This location is a British dependency composed of more than 300 islands in the North Atlantic Ocean.
2. About twenty of these islands are inhabited.
3. These islands are located about halfway between Nova Scotia and the West Indies.
4. Somerset Island is one of the islands in this group.

Name of country, province, or state: _____

Proof: _____

Creativity Across The Curriculum

1. Only small cars that can travel no faster than 20 mph are allowed here. Invent a machine that could be used for travel here that would require no gasoline and would not pollute.

2. Rainwater caught on the roofs of buildings is a leading source of fresh water here. Write a letter to your congressman expressing your concern about the effects of acid rain on our earth.

3. Using this week's newspaper, find as many articles as possible about different countries in the world. Classify them into groups of your own choice. (You must have at least three groups.)

Name _____

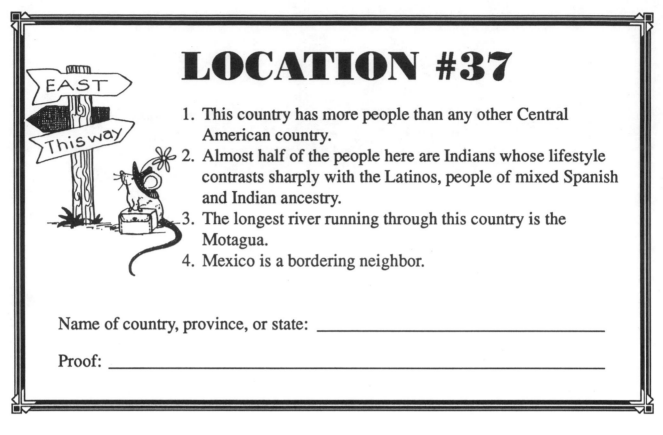

LOCATION #37

1. This country has more people than any other Central American country.
2. Almost half of the people here are Indians whose lifestyle contrasts sharply with the Latinos, people of mixed Spanish and Indian ancestry.
3. The longest river running through this country is the Motagua.
4. Mexico is a bordering neighbor.

Name of country, province, or state: _____

Proof: _____

Creativity Across The Curriculum

1. Using the key to a map, compute the distance from your school to the northern border of this country.

2. Cut out cartoon blocks from the comic strip sections of discarded newspapers. Assemble selected pictures to convey information about Location Number 37. Write original captions.

3. Make up a questionnaire that you could use to help determine where people would most like to go on a dream vacation and why they would like to go there.

Name _____

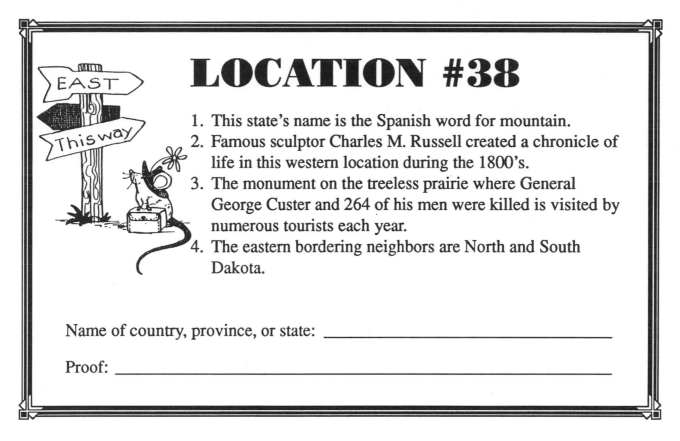

LOCATION #38

1. This state's name is the Spanish word for mountain.
2. Famous sculptor Charles M. Russell created a chronicle of life in this western location during the 1800's.
3. The monument on the treeless prairie where General George Custer and 264 of his men were killed is visited by numerous tourists each year.
4. The eastern bordering neighbors are North and South Dakota.

Name of country, province, or state: _____

Proof: _____

Creativity Across The Curriculum

1. Research the life of the Indian woman Sacajawea. Would she have been a good friend? Why or why not?

2. Buffalo once roamed this location by the thousands. Compare buffalo to the cows on modern-day farms. How are they different? Sketch each one.

3. Do some research to find out which Native American tribes lived in this state during the pioneer days. List the ways in which their lives changed once the land had been settled by Europeans.

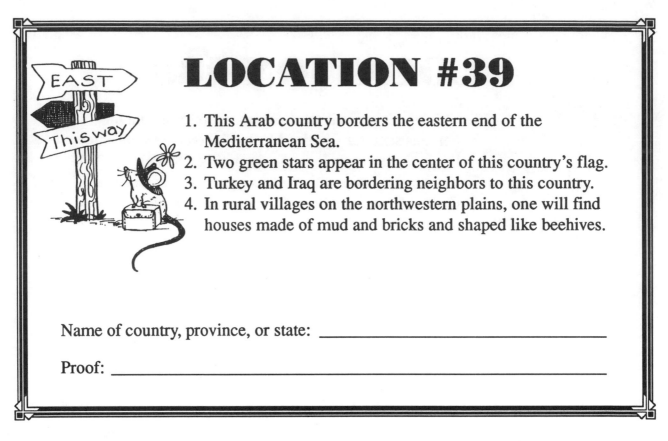

LOCATION #39

1. This Arab country borders the eastern end of the Mediterranean Sea.
2. Two green stars appear in the center of this country's flag.
3. Turkey and Iraq are bordering neighbors to this country.
4. In rural villages on the northwestern plains, one will find houses made of mud and bricks and shaped like beehives.

Name of country, province, or state: _____

Proof: _____

Creativity Across The Curriculum

1. In many Arabian countries, people cover their heads with various types of cloth as a matter of custom. If you were required to cover your head each day, what are five ways that you would do it?

2. Ancient irrigation systems using water wheels can still be found here. Compare a system using a water wheel to a modern-day irrigation system.

3. Due to its strategic location on a major trade route and its rich soil, this nation has been part of many battles and has been part of many nations. Create a drawing contrasting the present-day nation with the nation of the past.

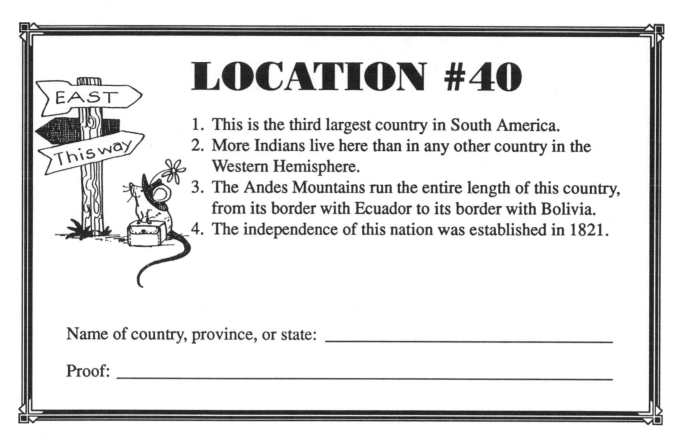

LOCATION #40

1. This is the third largest country in South America.
2. More Indians live here than in any other country in the Western Hemisphere.
3. The Andes Mountains run the entire length of this country, from its border with Ecuador to its border with Bolivia.
4. The independence of this nation was established in 1821.

Name of country, province, or state: _____

Proof: _____

Creativity Across The Curriculum

1. East of the Andes is a thick rain forest. Propose a law that would prevent this country from chopping down the rain forest for commercial use. The law must not endanger the economy of this nation. How would the law be enforced?

2. In order to limit the amount of timber destroyed each year, we must become more involved in recycling efforts. Find out how paper is recycled. Describe the process. What are you doing to help recycle paper?

3. The llama is used in this country to help transport goods through the rugged mountains. Research llamas. Create a flow chart to show a food chain of which this animal is a part.

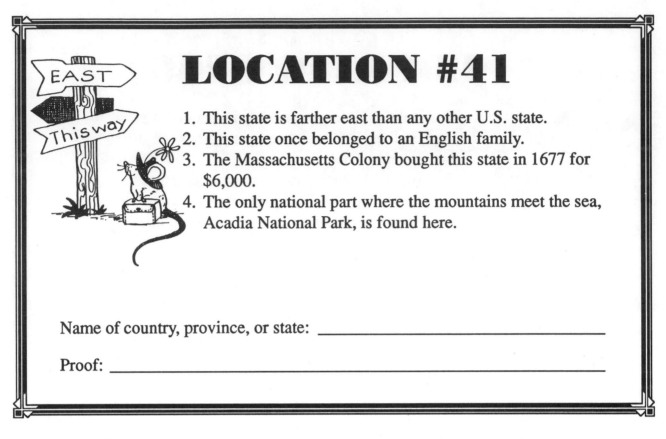

LOCATION #41

1. This state is farther east than any other U.S. state.
2. This state once belonged to an English family.
3. The Massachusetts Colony bought this state in 1677 for $6,000.
4. The only national part where the mountains meet the sea, Acadia National Park, is found here.

Name of country, province, or state: _____

Proof: _____

Creativity Across The Curriculum

1. If the U.S. wanted to add a fifty-first state, what would be the most logical area to create into a new state? What would you name it?

2. If a lobster off the coast of this state could talk, how would it describe a typical day here?

3. Brainstorm a list of famous people who lived in the U.S. during the 1700's.

Name _____

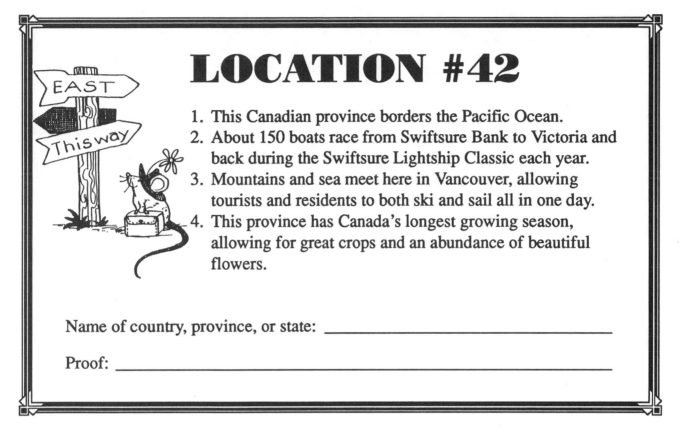

LOCATION #42

1. This Canadian province borders the Pacific Ocean.
2. About 150 boats race from Swiftsure Bank to Victoria and back during the Swiftsure Lightship Classic each year.
3. Mountains and sea meet here in Vancouver, allowing tourists and residents to both ski and sail all in one day.
4. This province has Canada's longest growing season, allowing for great crops and an abundance of beautiful flowers.

Name of country, province, or state: _____

Proof: _____

Creativity Across The Curriculum

1. Apple orchards abound here. List ten new and imaginative ways to use an apple.

2. A simile is a figure of speech that compares two things using the words "like" or "as." Create five similes that give you a better picture of life in this Canadian Province. For example: The mile-long Lion's Gate Bridge stretches like a bungee cord across Burrard Inlet, a gateway to the Pacific Ocean.

3. Write a newscaster's description of the weather here on July 1, Canada's National Birthday, and on Christmas Day. Be as descriptive and as accurate as possible.

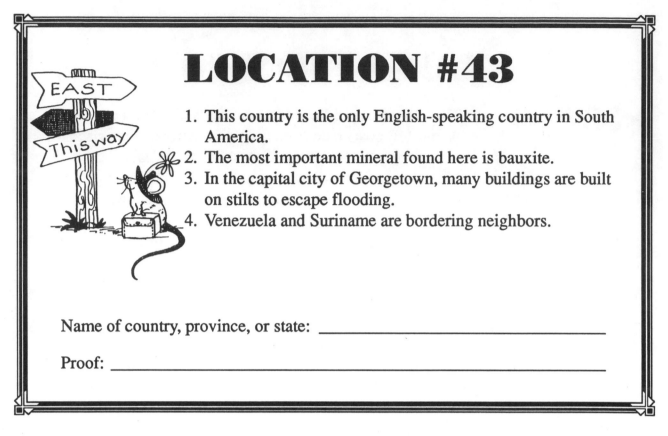

LOCATION #43

1. This country is the only English-speaking country in South America.
2. The most important mineral found here is bauxite.
3. In the capital city of Georgetown, many buildings are built on stilts to escape flooding.
4. Venezuela and Suriname are bordering neighbors.

Name of country, province, or state: _____

Proof: _____

Creativity Across The Curriculum

1. Why would other countries wish to purchase bauxite? How is it used?

2. Sugarcane is an important product of this country. We all know that sugary foods have little nutritional value. List five reasons why people continue to eat junk food. What do you think will happen to people who eat too much sugar?

3. Compare the size of this country to:
      Brazil
      Venezuela
      Peru
      Ecuador
  List them in order from largest to smallest.

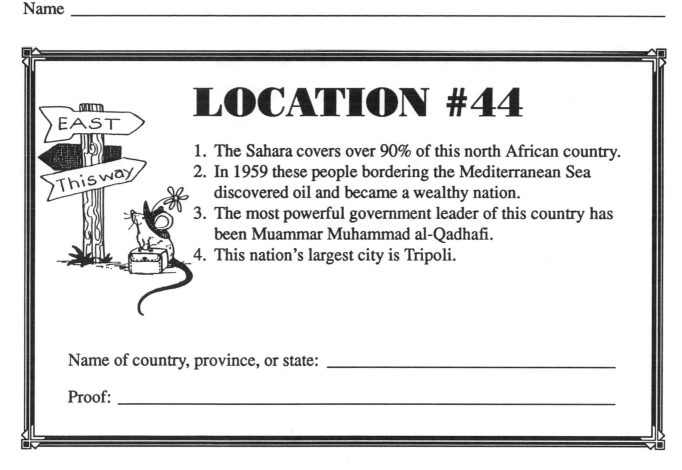

LOCATION #44

1. The Sahara covers over 90% of this north African country.
2. In 1959 these people bordering the Mediterranean Sea discovered oil and became a wealthy nation.
3. The most powerful government leader of this country has been Muammar Muhammad al-Qadhafi.
4. This nation's largest city is Tripoli.

Name of country, province, or state: _____

Proof: _____

Creativity Across The Curriculum

1. This nation, like many around the world, has experienced its share of conflict. Try to imagine a world without conflict, then respond to the following:
   • How would our world be different if there was no conflict of any type?
   • List as many kinds of conflict as possible.
   • What good can result from conflict?

2. What types of pygmy animals can you discover in this region? Research and write about one.

3. List five types of plants and animals that live in the Sahara Desert.

Name _____

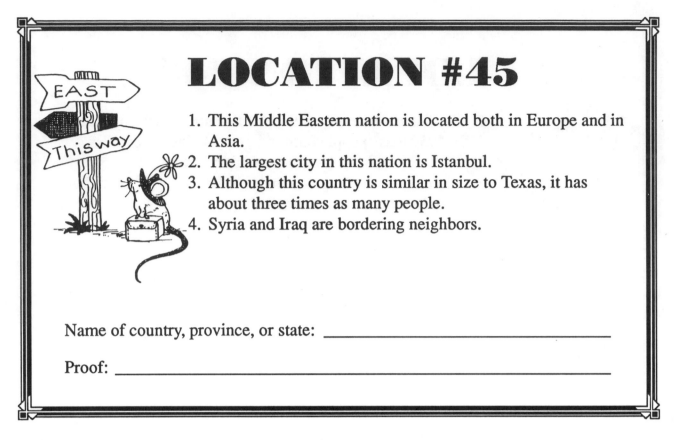

LOCATION #45

1. This Middle Eastern nation is located both in Europe and in Asia.
2. The largest city in this nation is Istanbul.
3. Although this country is similar in size to Texas, it has about three times as many people.
4. Syria and Iraq are bordering neighbors.

Name of country, province, or state: _____

Proof: _____

Creativity Across The Curriculum

1. Backgammon is a popular game in the coffee houses of this country. Learn to play it with a friend.

2. This country is famous for its classic Byzantine architecture. An example is the famous domed cathedral Hagia Sophia, built in 500 A.D. Handcrafted, beautifully-colored ceramic tiles decorate many palaces and mosques. Create your own tile pattern by making a geometric design that can be repeated over and over. This type of design is called a tessellation.

Name _____

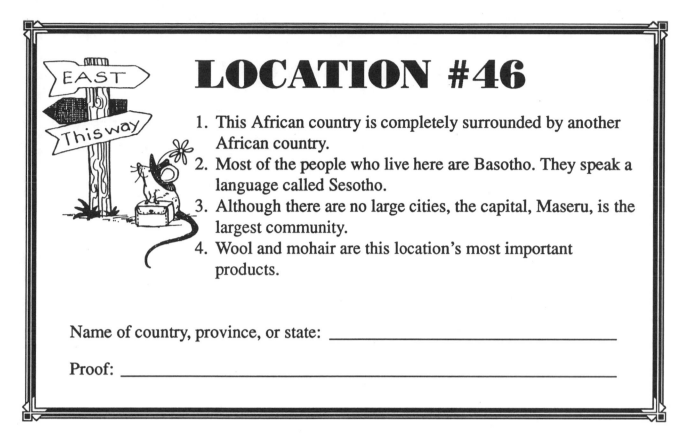

LOCATION #46

1. This African country is completely surrounded by another African country.
2. Most of the people who live here are Basotho. They speak a language called Sesotho.
3. Although there are no large cities, the capital, Maseru, is the largest community.
4. Wool and mohair are this location's most important products.

Name of country, province, or state: _____

Proof: _____

Creativity Across The Curriculum

1. A well-crafted, woven grass masterpiece, the Basotho hat, is the national symbol that appears on the country's flag. Develop a hat-sporting flag that symbolizes the U.S.

2. The Basotho house is a circular stone home with a thatched roof. Draw the floor plan for a circular house. What types of construction problems might occur?

3. Design a type of clothing or piece of furniture that could make use of mohair fabric. Sketch your design and indicate where the mohair would be used.

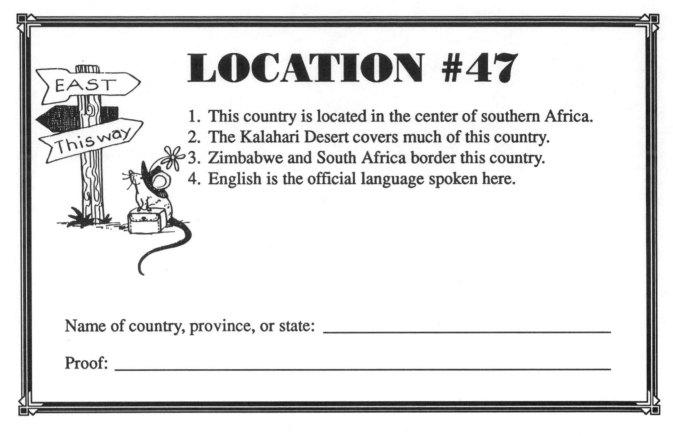

LOCATION #47

1. This country is located in the center of southern Africa.
2. The Kalahari Desert covers much of this country.
3. Zimbabwe and South Africa border this country.
4. English is the official language spoken here.

Name of country, province, or state: _____

Proof: _____

Creativity Across The Curriculum

1. About 10,000 Bushmen live in this country, gathering food and hunting in the Kalahari. If you could give these primitive people one modern convenience, what would it be and why?

2. The continent of Africa has suffered much racial tension. Do some research to find the countries where black Africans are allowed to become rulers or political leaders.

3. Compare the population of this country to the population of your state. How many people per square mile are there in each area?

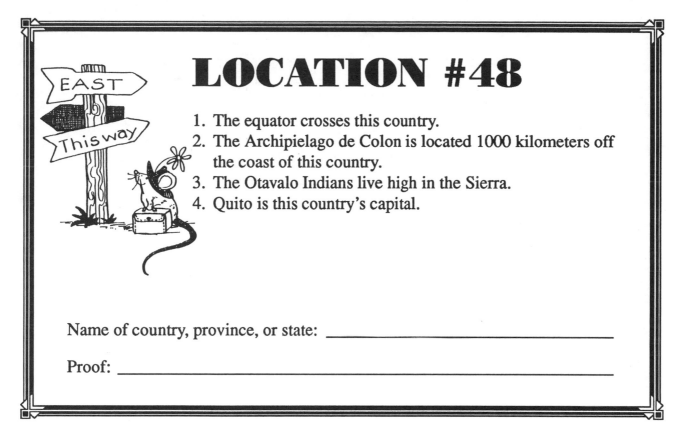

LOCATION #48

1. The equator crosses this country.
2. The Archipielago de Colon is located 1000 kilometers off the coast of this country.
3. The Otavalo Indians live high in the Sierra.
4. Quito is this country's capital.

Name of country, province, or state: _____

Proof: _____

Creativity Across The Curriculum

1. Fifty percent of the homes in this country are made of adobe and reed, cane, or other scrap materials. If you could not use lumber, metal, glass, or bricks, how would you build a house? Draw your design and label your materials.

2. Guinea pigs are raised in this country as a source of food. If we no longer had pork, beef, chicken, turkey, fish, or other seafood around, what would you choose to eat?

3. Looms are found in many homes in this country. Create and build a simple loom of your own with a piece of cardboard and some string.

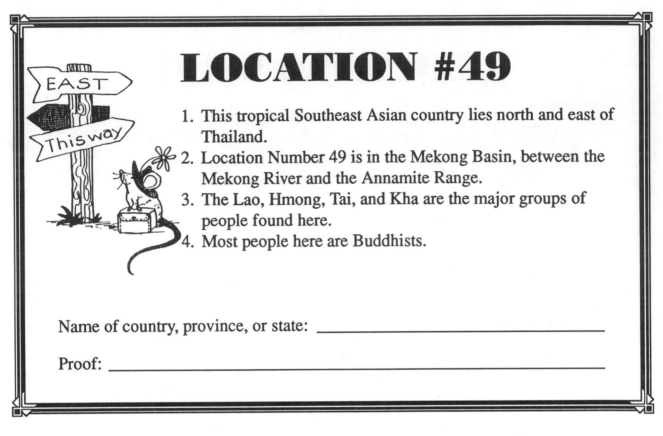

LOCATION #49

1. This tropical Southeast Asian country lies north and east of Thailand.
2. Location Number 49 is in the Mekong Basin, between the Mekong River and the Annamite Range.
3. The Lao, Hmong, Tai, and Kha are the major groups of people found here.
4. Most people here are Buddhists.

Name of country, province, or state: _____

Proof: _____

Creativity Across The Curriculum

1. Old-fashioned farming equipment keeps the economy of this country from improving. If you wanted to aid this country in its war on poverty, what are some changes that you would recommend? Taking into account the culture of this nation, what ten suggestions could you make?

2. The Hmong women create lovely, ornate needlework. Create an ornate pattern for a small quilt block. Color it using only primary colors and black.

LOCATION #50

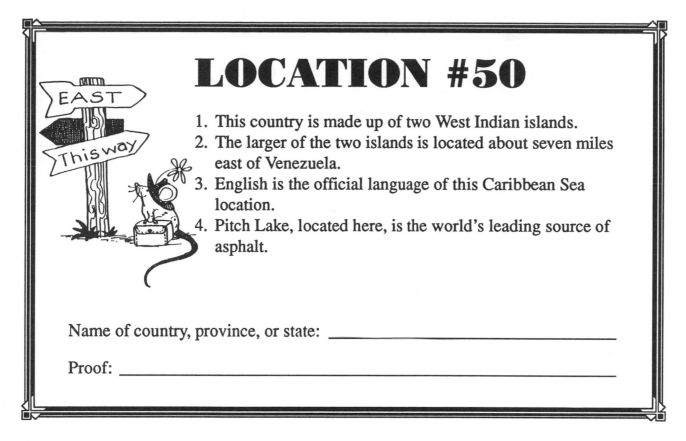

1. This country is made up of two West Indian islands.
2. The larger of the two islands is located about seven miles east of Venezuela.
3. English is the official language of this Caribbean Sea location.
4. Pitch Lake, located here, is the world's leading source of asphalt.

Name of country, province, or state: _____

Proof: _____

Creativity Across The Curriculum

1. This country is the home of calypso music. Compare calypso to rock and roll. How are they the same? Different?

2. The limbo dance originated here. With a couple of friends, see how well you can do the limbo.

3. Many people here play "pans," native musical instruments made from empty oil drums. Create a small cylinder drum of some type and learn to play basic rhythms with it.

Name _____

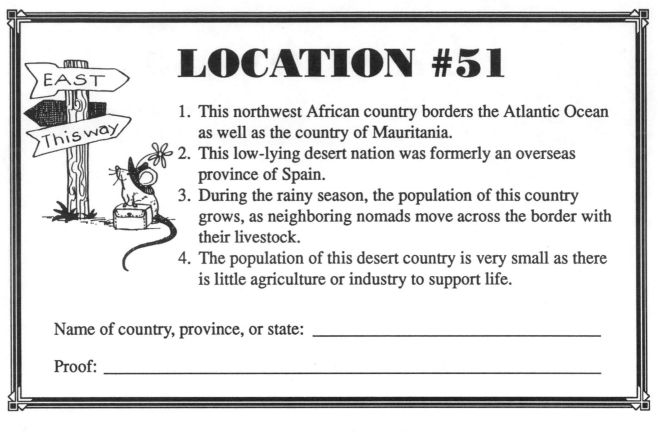

LOCATION #51

1. This northwest African country borders the Atlantic Ocean as well as the country of Mauritania.
2. This low-lying desert nation was formerly an overseas province of Spain.
3. During the rainy season, the population of this country grows, as neighboring nomads move across the border with their livestock.
4. The population of this desert country is very small as there is little agriculture or industry to support life.

Name of country, province, or state: _____

Proof: _____

Creativity Across The Curriculum

1. Across this dusty desert nation, caravans of camels and riders journey, searching for water. Create a journal entry for a typical day in the life of one of these riders.

2. Tomorrow you will be asked to sit down with the political leaders of the country of Africa. What advice will you give them and why? Write out the speech you will deliver to them.

3. Crime is rising in this small country as it is all over the world. What do you believe to be the reasons for this increase?

Name _____

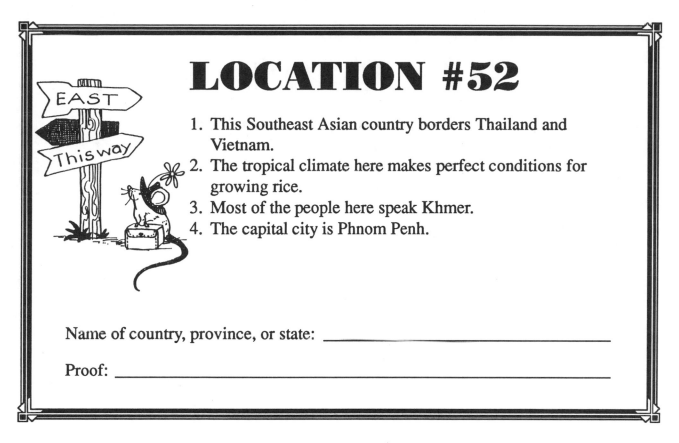

LOCATION #52

1. This Southeast Asian country borders Thailand and Vietnam.
2. The tropical climate here makes perfect conditions for growing rice.
3. Most of the people here speak Khmer.
4. The capital city is Phnom Penh.

Name of country, province, or state: _____

Proof: _____

Creativity Across The Curriculum

1. Rice and fish are the main foods of the people living in this country. If you had to survive and remain healthy for many years, what three additional foods would you add to this diet?

2. The sun during the day here is extremely intense as people work in the fields. For years they have worn a hat that is somewhat triangular in shape and extremely plain. It is made of straw-like fibers. Create a new, more up-to-date, yet functional hat. Sketch and color it. Create a paper prototype.

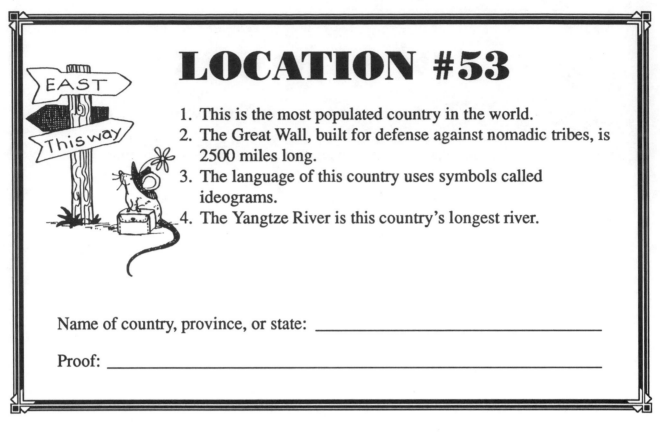

LOCATION #53

1. This is the most populated country in the world.
2. The Great Wall, built for defense against nomadic tribes, is 2500 miles long.
3. The language of this country uses symbols called ideograms.
4. The Yangtze River is this country's longest river.

Name of country, province, or state: _____

Proof: _____

Creativity Across The Curriculum

1. The most respected philosopher in the history of this country was Confucius. His ideas stress developing moral character and responsibility. Find one of Confucius' principles. Rewrite it using modern language.

2. If you wished to travel the length of the Great Wall, 2,500 miles, and could walk 15 miles in a day, how long would it take you to get from one end to the other in months and days? In weeks and days? In hours?

3. If Confucius and Michelangelo were both alive today, who would you choose to be your best friend? Support your choice.

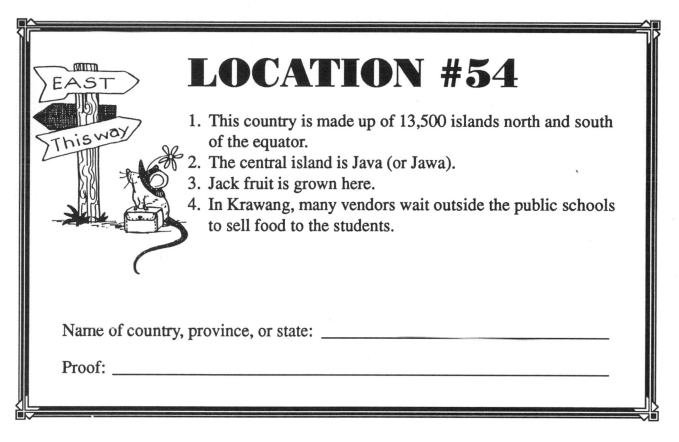

LOCATION #54

1. This country is made up of 13,500 islands north and south of the equator.
2. The central island is Java (or Jawa).
3. Jack fruit is grown here.
4. In Krawang, many vendors wait outside the public schools to sell food to the students.

Name of country, province, or state: _____

Proof: _____

Creativity Across The Curriculum

1. There is only one television channel on the island of Java. If you only had one television station available what programs would you want to be sure to include? Create a programming schedule that would run from 8:00 a.m. to 10:30 p.m.

2. One of the performing arts of this country is "mask dancing." These dances are based on Indian and Javanese tales. With a friend, create a set of masks that you can use while telling a folktale. Practice before giving your presentation to the class.

3. The art of batik began here. Try your hand at it. Begin with white cloth. Sketch a design on the cloth with light pencil, then brush over it with melted paraffin. (The paraffin, or wax, may be melted and held at a constant temperature in an old electric frying-pan, or melted in a clean, empty can placed in a saucepan of simmering water.) Dye your cloth with a cold water dye, following instructions on the package. Remove the wax by scraping or boiling the cloth, or ironing it between newspapers with an old iron. Sketch and wax a new design. Dye with a second color. Continue the process until the desired color scheme and design have been achieved.

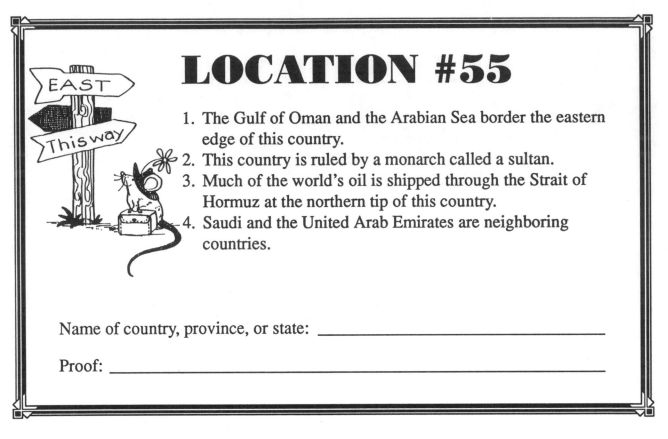

LOCATION #55

1. The Gulf of Oman and the Arabian Sea border the eastern edge of this country.
2. This country is ruled by a monarch called a sultan.
3. Much of the world's oil is shipped through the Strait of Hormuz at the northern tip of this country.
4. Saudi and the United Arab Emirates are neighboring countries.

Name of country, province, or state: _____

Proof: _____

Creativity Across The Curriculum

1. Oil was discovered in this country in 1964. Until that time, the barren wasteland covering much of the nation caused it to remain poor and undeveloped. When Sultan Qaboos took over in 1970, he developed the oil industry, built roads, hospitals and schools, and promoted irrigation. Until 1970, few girls in this country ever attended school. Why would a country like this be more likely to have let boys rather than girls attend school?

2. Many of the rural men here wear turbans on their heads to protect themselves from the summer sun. Using a strip of old cloth, create a turban that would protect your head from the sun. Write out the directions for your creation so that someone else may reproduce the same type of turban.

3. This country has no constitution. If you could have only ten rules by which to govern an entire country, what would those rules be? List them in order of importance.

Name _____

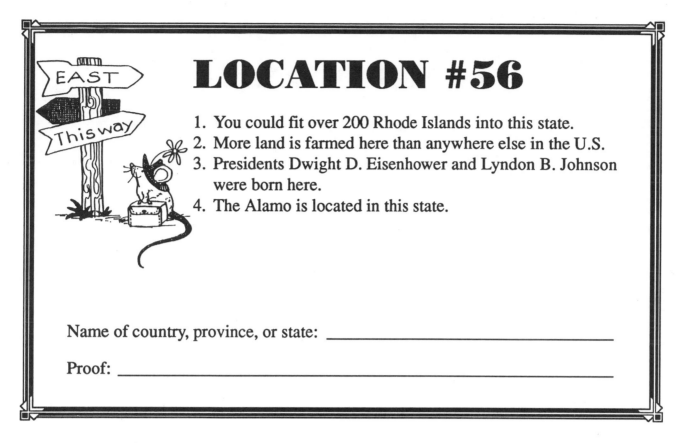

LOCATION #56

1. You could fit over 200 Rhode Islands into this state.
2. More land is farmed here than anywhere else in the U.S.
3. Presidents Dwight D. Eisenhower and Lyndon B. Johnson were born here.
4. The Alamo is located in this state.

Name of country, province, or state: _____

Proof: _____

Creativity Across The Curriculum

1. Using only egg cartons, glue, paper, and toothpicks, recreate a scale model of the Alamo. Find out how Davy Crockett and Jim Bowie were involved with the Alamo.

2. Draw a picture of the state seal. Do some research to find out the significance of the symbols on the seal.

3. List five important events that occurred during the presidency of Dwight D. Eisenhower, and five that occurred during the presidency of Lyndon B. Johnson. Which president do you feel accomplished more for our country? Why?

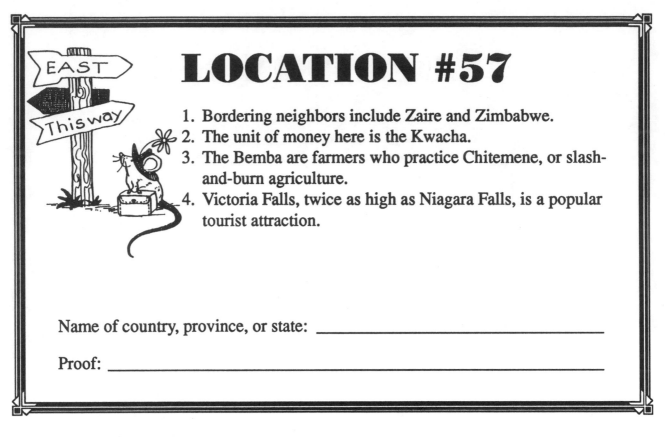

LOCATION #57

1. Bordering neighbors include Zaire and Zimbabwe.
2. The unit of money here is the Kwacha.
3. The Bemba are farmers who practice Chitemene, or slash-and-burn agriculture.
4. Victoria Falls, twice as high as Niagara Falls, is a popular tourist attraction.

Name of country, province, or state: _____

Proof: _____

Creativity Across The Curriculum

1. When the Bemba slash-and-burn brush on the land, they mix the ash with poor soil to use as a type of fertilizer. After three to four years of using this process, the Bemba move on. If you could go to this country and teach the farmers improved farming techniques, what advice would you give them? Design a brochure that you could give to African farmers to help them improve their yield.

2. This country is one of the world's leading producers of copper. Brainstorm a list of uses for this mineral.

3. Using a rebus format, create a paragraph of information about Location Number 57. Instead of using only words to convey your message, use a combination of words and pictures. The pictures may be cut from magazines or hand-drawn.

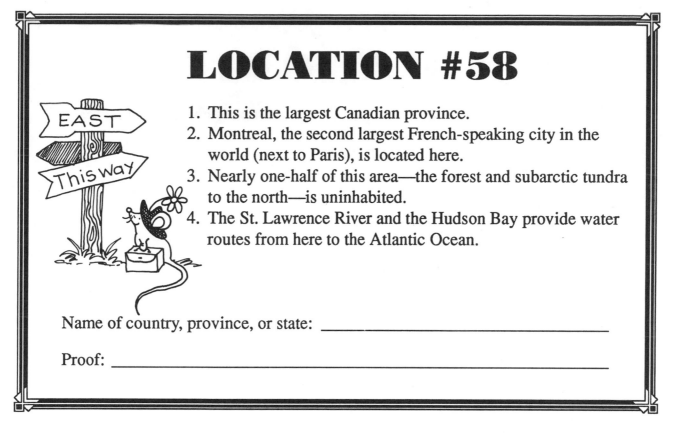

LOCATION #58

1. This is the largest Canadian province.
2. Montreal, the second largest French-speaking city in the world (next to Paris), is located here.
3. Nearly one-half of this area—the forest and subarctic tundra to the north—is uninhabited.
4. The St. Lawrence River and the Hudson Bay provide water routes from here to the Atlantic Ocean.

Name of country, province, or state: _____

Proof: _____

Creativity Across The Curriculum

1. Though three-fourths of the population of Canada speaks English, the official language of this province is French. All road signs, as well as most menus, public displays, and news information, are written in French. Write a letter to the Canadian Prime Minister asking him if he thinks it important to retain French as the official language for this province. Tell him your opinions.

2. The stores that line the waterfront of the St. Lawrence River in Montreal may be different from the kinds of stores in your town. Create a list of stores that you would expect to see along the waterfront.

3. Go to the library and find a book about Canada. Look at the illustrations. Create a list of 25 adjectives that describe the pictures you have seen.

Name _____

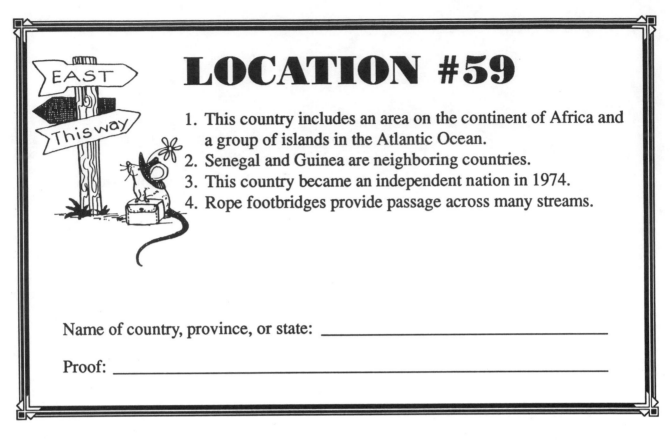

LOCATION #59

1. This country includes an area on the continent of Africa and a group of islands in the Atlantic Ocean.
2. Senegal and Guinea are neighboring countries.
3. This country became an independent nation in 1974.
4. Rope footbridges provide passage across many streams.

Name of country, province, or state: _____

Proof: _____

Creativity Across The Curriculum

1. The total land area of this country is about 14,000 square miles. List the countries of Africa from largest to smallest in land area.

2. Using the latest population figures, arrange the countries of Africa from most populated to least populated.

3. Create an ABC Poem to describe the African continent. For example:

Africa, Big, Conakry
Drought, Elephants, Fulani
Guinea…

LOCATION #60

EAST
This way

1. The Lake Pontchartrain Causeway located here is the world's longest bridge.
2. Many fur trappers and fishermen live and work in the marshes here.
3. Jean Laffite National Historical Park commemorates the last battle of the War of 1812.
4. Texas borders this state.

Name of country, province, or state: _____

Proof: _____

Creativity Across The Curriculum

1. Mardi Gras is a nationally known celebration held yearly in this state. Create a costume that you could wear to highlight your heritage. Sketch it and describe its significance.

2. Use alliteration to create five facts concerning this state. For example: Mighty Mississippi, museums, and marshes make a masterpiece.

3. This state is a leading producer of salt. Create a cartoon explaining the importance of salt.

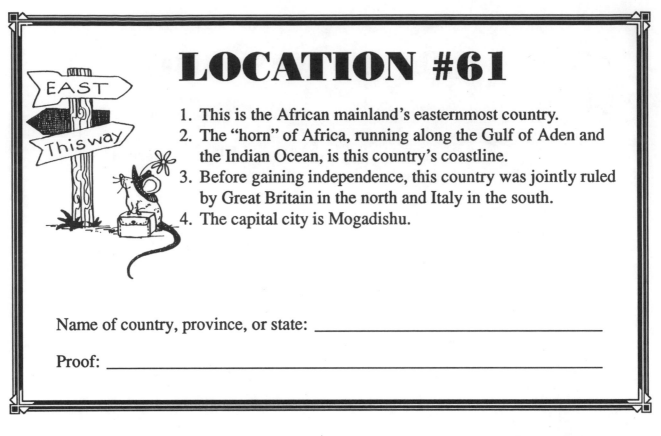

LOCATION #61

1. This is the African mainland's easternmost country.
2. The "horn" of Africa, running along the Gulf of Aden and the Indian Ocean, is this country's coastline.
3. Before gaining independence, this country was jointly ruled by Great Britain in the north and Italy in the south.
4. The capital city is Mogadishu.

Name of country, province, or state: _____

Proof: _____

Creativity Across The Curriculum

1. Most of this country's people belong to one of four clans. What part has clan loyalty played in literature? Research Shakespeare's "Romeo and Juliet" as an example.

2. Poetry and singing are two favorite types of entertainment in Location Number 61. Songs often tell of love, death, or war. What are the most often repeated themes of popular songs in the United States today? How do these themes reflect our culture? List the major themes and examples of titles promoting each one.

Name _____

LOCATION #62

1. This country is located on the Balkan Peninsula, and is neighbor to Austria and Hungary.
2. The most important river here is the Danube.
3. One of this country's six republics, Slovenia, grows murka, a plant with dark brown flowers that smell like fine milk chocolate.
4. Mother Teresa was born here in the Macedonian city of Skopje.

Name of country, province, or state: _____

Proof: _____

Creativity Across The Curriculum

1. When Austria ruled Slovenia, the people were not allowed to speak their own language or to teach it. Valentin Vodnik began work on a Slovenian dictionary, but Austrian officials would not allow him to finish it. If you could capture in print the twenty-five most important words of the English language before English was forgotten completely, which words would you choose?

2. Find one important accomplishment of each of the following famous people.
   • Thomas Edison
   • Michelangelo
   • Johnny Carson
   • Mother Teresa
   • Dr. Martin Luther King, Jr.
   • Jonas Salk

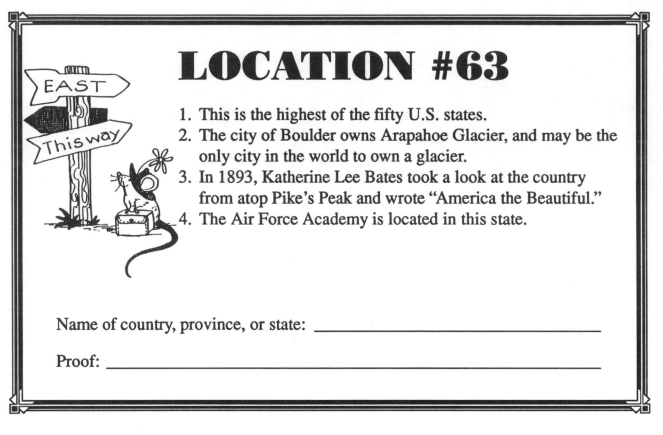

LOCATION #63

1. This is the highest of the fifty U.S. states.
2. The city of Boulder owns Arapahoe Glacier, and may be the only city in the world to own a glacier.
3. In 1893, Katherine Lee Bates took a look at the country from atop Pike's Peak and wrote "America the Beautiful."
4. The Air Force Academy is located in this state.

Name of country, province, or state: _____

Proof: _____

Creativity Across The Curriculum

1. The world's highest suspension bridge goes across this state's Royal Gorge. The bridge is suspended 1,053 feet above the base of the gorge. Do some research to find out how a suspension bridge is built. Explain the process to your class. Use diagrams or charts to aid in your explanation.

2. There are 5,280 feet in a mile. Construct five math problems using factual information about heights in this state. Make story problems that involve at least two different arithmetic processes in each.

3. Skiing is an important sport for the people who live here, as well as for the tourists who visit. Do some research to find out how the sport of skiing originated. Create a simple non-fiction booklet describing the way in which skiing began and developed.

LOCATION #64

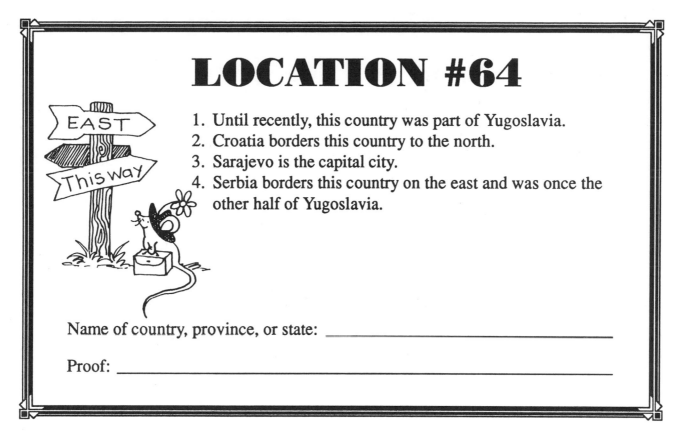

1. Until recently, this country was part of Yugoslavia.
2. Croatia borders this country to the north.
3. Sarajevo is the capital city.
4. Serbia borders this country on the east and was once the other half of Yugoslavia.

Name of country, province, or state: _____

Proof: _____

Creativity Across The Curriculum

1. When this country broke away from Yugoslavia in 1992, it had to create a new flag. If you were in charge of creating a flag for this new country, what would it look like? Sketch your flag and color it. Explain the reason for your design.

2. Describe Location Number 64 through the "eyes" of a tourist's suitcase.

3. The old flag of Yugoslavia was composed entirely of geometric shapes. Create a new geometric shape and give it a name. Draw it.

LOCATION #65

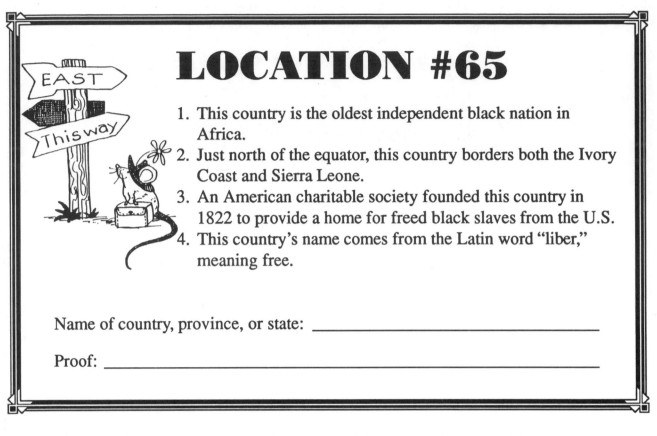

1. This country is the oldest independent black nation in Africa.
2. Just north of the equator, this country borders both the Ivory Coast and Sierra Leone.
3. An American charitable society founded this country in 1822 to provide a home for freed black slaves from the U.S.
4. This country's name comes from the Latin word "liber," meaning free.

Name of country, province, or state: _____

Proof: _____

Creativity Across The Curriculum

1. Ninety-five percent of the people here are native Africans of 16 ethnic groups. Each group speaks a different language and observes different customs. A few of these ethnic groups are the Kpelles, the Lomas, the Bassa, and the Kru. Do some research to discover the names of as many of the 16 groups as possible. Compare the lifestyle of one of these groups to your style of living.

2. When President Lincoln made the decision to free the slaves, Americans were not all in agreement. Today, other issues of great importance have surfaced in America. Once again, Americans are not all in agreement. List five key issues in America on which the people are divided. How do you feel we can come to some kind of agreement on these issues, and move forward as a nation united?

Mini-Dictionary

Create a mini-dictionary with definitions and illustrations for each of the following words of importance to this country.

cassava	duikers
harmattan	indigenous
Monrovia	pygmy hippopotamus
taro	winnow

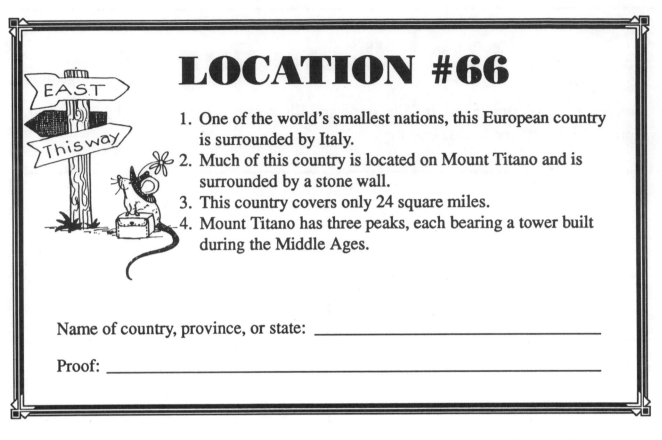

LOCATION #66

1. One of the world's smallest nations, this European country is surrounded by Italy.
2. Much of this country is located on Mount Titano and is surrounded by a stone wall.
3. This country covers only 24 square miles.
4. Mount Titano has three peaks, each bearing a tower built during the Middle Ages.

Name of country, province, or state: _____

Proof: _____

Creativity Across The Curriculum

1. The beautiful stamps produced in this country are sought by stamp collectors from around the world. Design a new stamp that would depict an aspect of this country's history.

2. This country is a favorite vacation spot. Take a survey to determine the favorite vacation spots of the members of your class. Present your findings in graph form.

3. Pick a cartoon character. Create an adventure for this character on Mount Titano. Draw the cartoon adventure.

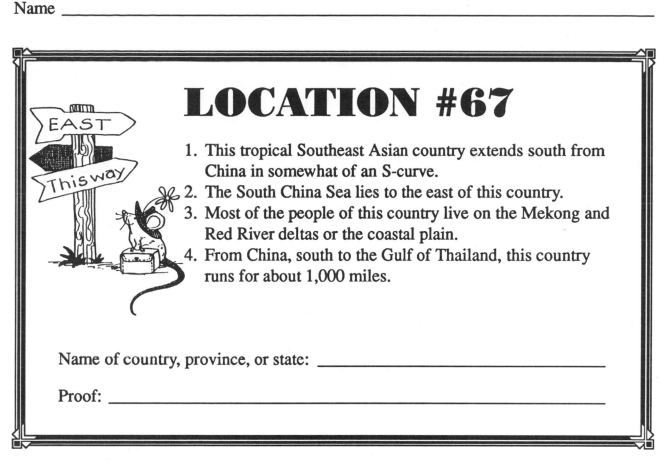

LOCATION #67

1. This tropical Southeast Asian country extends south from China in somewhat of an S-curve.
2. The South China Sea lies to the east of this country.
3. Most of the people of this country live on the Mekong and Red River deltas or the coastal plain.
4. From China, south to the Gulf of Thailand, this country runs for about 1,000 miles.

Name of country, province, or state: _____

Proof: _____

Creativity Across The Curriculum

1. This country has been involved in war for most of its existence since the 1800's. List the conflicts in order of their occurrence and the outcome of each conflict. What has been the major reason for such continual strife?

2. Poetry is the most important type of literature in this country. Create some "Definition Poetry" about this nation. Definition Poetry involves taking a topic and describing it using short, creative, and informative phrases. Example:

> Mekong Delta
> Formed by swiftly flowing rivers
> Rich, swampy soil
> Rice fields abound.
> People, people everywhere
> Planting, harvesting, living,
> Wishing for peace.

3. This country has one of the world's largest armed forces. Do some research to find five facts about each of our U.S. branches of the armed forces. Which branch do you prefer and why?

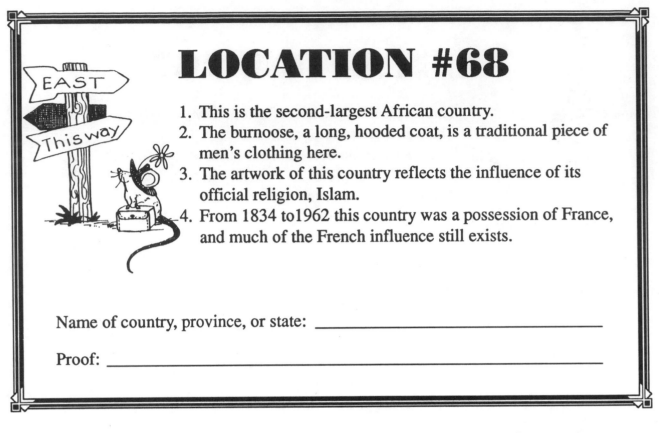

LOCATION #68

1. This is the second-largest African country.
2. The burnoose, a long, hooded coat, is a traditional piece of men's clothing here.
3. The artwork of this country reflects the influence of its official religion, Islam.
4. From 1834 to1962 this country was a possession of France, and much of the French influence still exists.

Name of country, province, or state: _____

Proof: _____

Creativity Across The Curriculum

1. A hot desert, the Sahara, covers four-fifths of this country. Detail the ways in which your life would change if you lived in the Sahara.

2. In the summer, the sirocco can cause distress for people traveling. Since "One picture may be worth a thousand words," draw your depiction of a sirocco.

3. Camel caravans are still used here to cross the Sahara. Prepare a story and picture about a camel.

Name _____

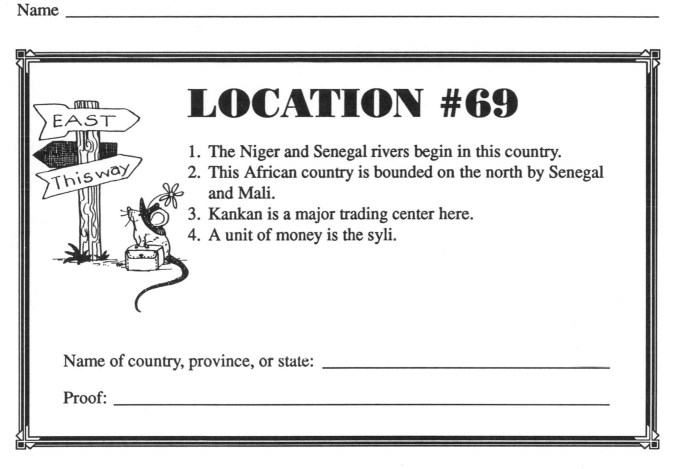

LOCATION #69

1. The Niger and Senegal rivers begin in this country.
2. This African country is bounded on the north by Senegal and Mali.
3. Kankan is a major trading center here.
4. A unit of money is the syli.

Name of country, province, or state: _____

Proof: _____

Creativity Across The Curriculum

1. The history of this country is handed down orally from generation to generation by storytellers called griots. Prepare a story that a griot might share about U.S. history.

2. Folktales are an important part of this nation. Create a puppet to tell a folktale about some aspect of this culture.

3. If you were one of the writers for the game show Jeopardy, what are two questions you would ask about this country? Try them out on a couple of friends.

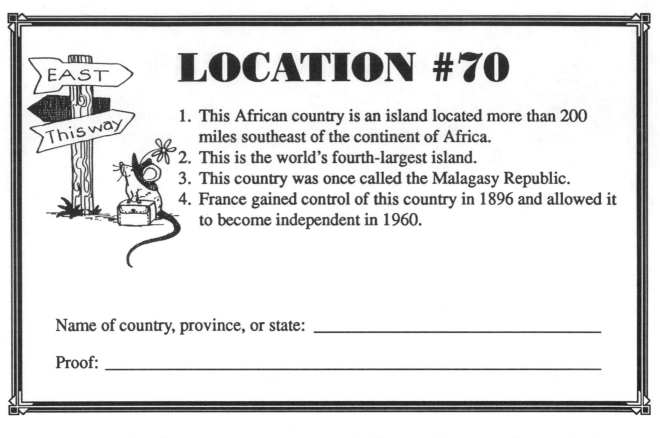

LOCATION #70

1. This African country is an island located more than 200 miles southeast of the continent of Africa.
2. This is the world's fourth-largest island.
3. This country was once called the Malagasy Republic.
4. France gained control of this country in 1896 and allowed it to become independent in 1960.

Name of country, province, or state: _____

Proof: _____

Creativity Across The Curriculum

1. The lemur is found only in this country and on the Comoros Islands. Create a diorama of the lemur's natural habitat. Give a presentation to a small group, or your class, on this interesting animal.

2. Create a large outline of the world on butcher paper. Using magazine and newspaper pictures, as well as sketches, create a global vision of transportation around the world. Next to each picture write the name of the country, state, or province where such transportation might be found.

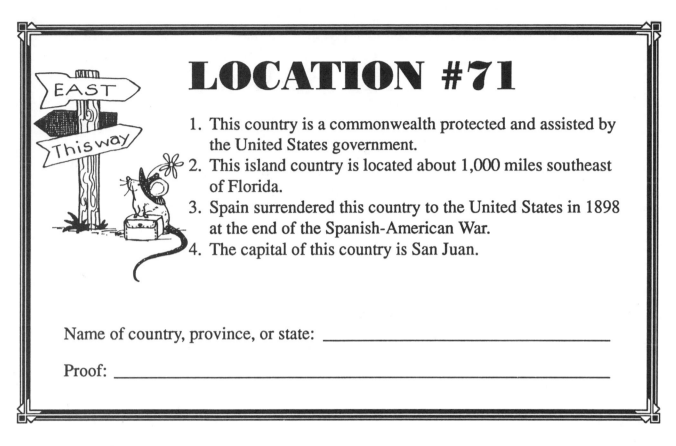

LOCATION #71

1. This country is a commonwealth protected and assisted by the United States government.
2. This island country is located about 1,000 miles southeast of Florida.
3. Spain surrendered this country to the United States in 1898 at the end of the Spanish-American War.
4. The capital of this country is San Juan.

Name of country, province, or state: _____

Proof: _____

Creativity Across The Curriculum

1. Some of the trees here produce fruits and nuts not grown elsewhere in the U.S. Examples are breadfruit, guanabanas, sea grapes, papayas, and star apples. Do some research to find out more about each one. Which one do you think would be your favorite? Why?

2. One of this location's holidays is called Three Kings' Day. It is celebrated on January 6, marking the end of Christmas. On this day, as at Christmas, children receive presents. Prepare a letter for the President of the United States convincing him to declare Three Kings' Day a national holiday.

3. Christopher Columbus is believed to have landed here in 1493. Do some research to find out what part Columbus played in the history of the United States.

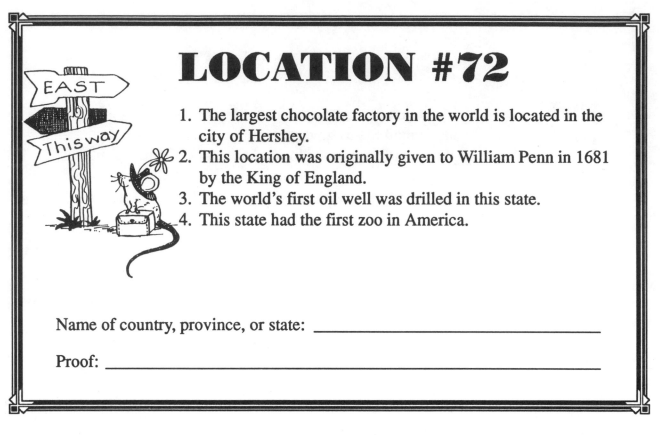

LOCATION #72

1. The largest chocolate factory in the world is located in the city of Hershey.
2. This location was originally given to William Penn in 1681 by the King of England.
3. The world's first oil well was drilled in this state.
4. This state had the first zoo in America.

Name of country, province, or state: _____

Proof: _____

Creativity Across The Curriculum

1. The Declaration of Independence was signed here at Independence Hall. Write a modern-day version of the Declaration.

2. The first television broadcast originated here. If you could have your own television show for one month, one-half hour per week, what type of show would you create?

3. American teens danced for many years to the popular music played on American Bandstand, broadcast from this location. With a partner, create a dance that could be featured on a teen dance show.

Name _____

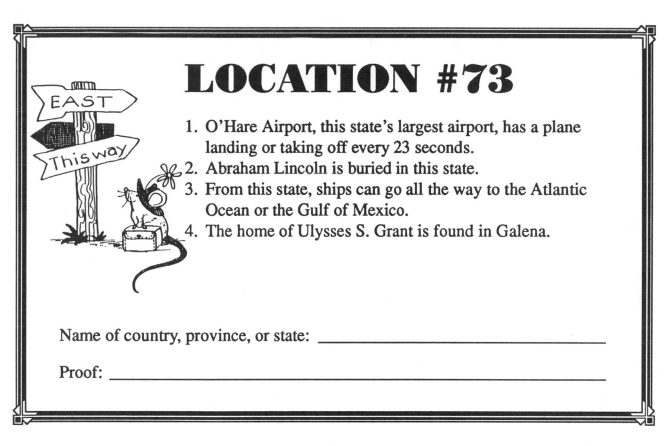

LOCATION #73

1. O'Hare Airport, this state's largest airport, has a plane landing or taking off every 23 seconds.
2. Abraham Lincoln is buried in this state.
3. From this state, ships can go all the way to the Atlantic Ocean or the Gulf of Mexico.
4. The home of Ulysses S. Grant is found in Galena.

Name of country, province, or state: _____

Proof: _____

Creativity Across The Curriculum

1. Among the exhibits at the Museum of Science and Industry are a coal mine, a submarine, and a space capsule. If the museum would allow your class to decide what the next exhibit should be, what would they choose? Survey your class for ideas. Then write a letter to the president of the museum explaining the display ideas that your class would like to suggest.

2. The Shedd Aquarium, found here, is one of the most beautiful displays of live fish in the U.S. Create an interesting and unique "aquarium in a shoebox." Choose any materials that you would like to use to create your lifelike fish and lake or sea life. Suspend them from the top of the box with fish line. When your aquarium is finished, cover it with plastic wrap.

LOCATION #74

1. The capital of this Canadian province is Edmonton.
2. Petroleum is an important product produced in this province.
3. British Columbia borders this province on the west.
4. The Calgary Stampede, a world-famous rodeo, is held here each year.

Name of country, province, or state: _____

Proof: _____

Creativity Across The Curriculum

1. Create a map of all of the Canadian provinces. Pinpoint the places where early tribes of Indians settled. Make a key for your map so that the various tribes of Indians may be easily recognized.

2. Has the expansion of the petroleum industry created problems for the wildlife of this province? Make a list of endangered animals. Do any of them live here, or have any of them lived here in the past?

3. Write a convincing argument for or against individuals or businesses using an area's resources for profitable gains. Debate the issue with a friend who may not agree with your position.

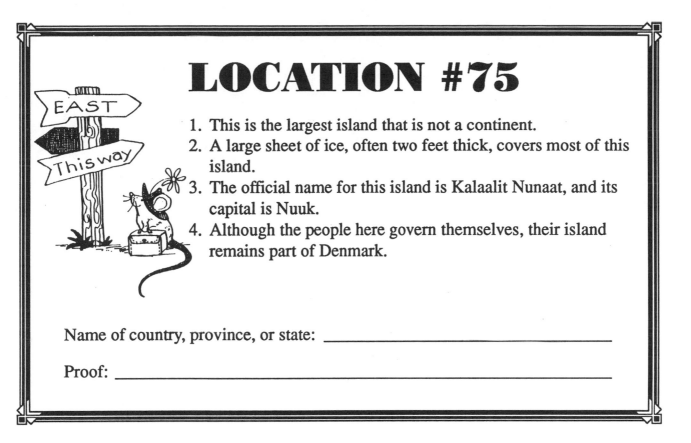

LOCATION #75

1. This is the largest island that is not a continent.
2. A large sheet of ice, often two feet thick, covers most of this island.
3. The official name for this island is Kalaalit Nunaat, and its capital is Nuuk.
4. Although the people here govern themselves, their island remains part of Denmark.

Name of country, province, or state: _____

Proof: _____

Creativity Across The Curriculum

1. This country's name is a compound word. In four minutes, list as many compound words as you can. Examples:

    1. bulldog   4. cowgirl
    2. football   5. mailmen
    3. pigpen

BONUS: What are two other countries whose names are compound words?

2. The Inuit people fish and raise sheep and reindeer. Describe these people.

3. Erik the Red falsely named this country to lure settlers from Iceland. The settlers must have been angry to find all the ice and cold. Create a wanted poster for the capture of Erik the Red. Make it as factual as possible.

4. Create an ad designed to increase the population of this country by luring people to its shores.

Name _____

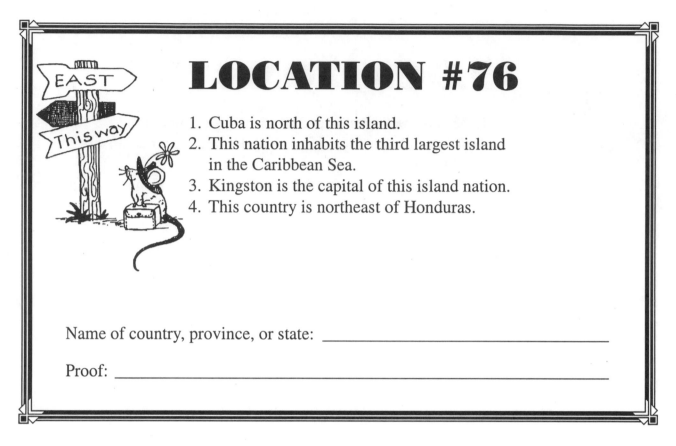

LOCATION #76

1. Cuba is north of this island.
2. This nation inhabits the third largest island in the Caribbean Sea.
3. Kingston is the capital of this island nation.
4. This country is northeast of Honduras.

Name of country, province, or state: _____

Proof: _____

Creativity Across The Curriculum

1. Create a travel poster that verbally and visually lures tourists to the shores of this Caribbean Island.

2. Three-fourths of the people in this country have descended from African slaves who were plantation workers during the time of British colonial rule. Create a pictorial time line which shows the changes that have occurred in the lives of these people during the last 200 years.

3. Aliens have just landed on this Caribbean island. Brainstorm ideas about how the aliens might use each of the following items if they had never seen them before.

- snorkel
- surfboard
- tires
- canoe
- lipstick
- sunglasses

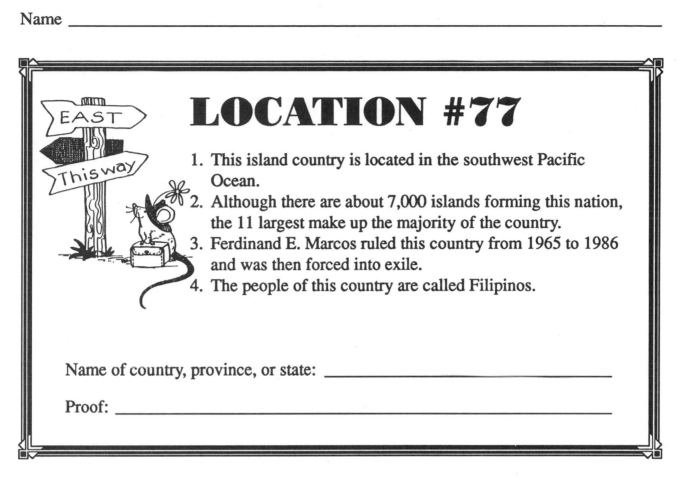

LOCATION #77

1. This island country is located in the southwest Pacific Ocean.
2. Although there are about 7,000 islands forming this nation, the 11 largest make up the majority of the country.
3. Ferdinand E. Marcos ruled this country from 1965 to 1986 and was then forced into exile.
4. The people of this country are called Filipinos.

Name of country, province, or state: _____

Proof: _____

Creativity Across The Curriculum

1. Tarsiers live only in this country and the East Indies. Sketch one and describe its important features.

2. Vehicles called jeepneys provide much of the local transportation here. These vehicles are like taxis, shared by as many people as possible at one time. Few people have their own cars. Design a "jeepney" that could be used successfully in crowded U.S. cities.

3. An average of five typhoons hit this country each year—damaging property and killing and injuring people. Compare a typhoon, a hurricane, and a tornado. How are they the same? How are they different?

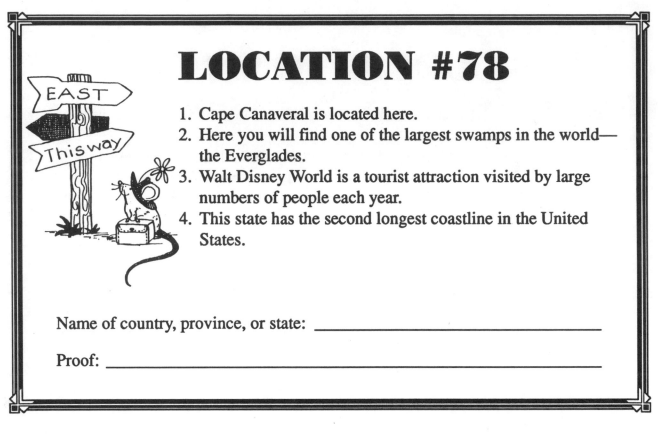

LOCATION #78

1. Cape Canaveral is located here.
2. Here you will find one of the largest swamps in the world—the Everglades.
3. Walt Disney World is a tourist attraction visited by large numbers of people each year.
4. This state has the second longest coastline in the United States.

Name of country, province, or state: _____

Proof: _____

Creativity Across The Curriculum

1. Years of planning and design transpired for Disney World/Epcot Center/MGM to become a reality. If you had been part of the planning committee, what would you have wanted to include? Draw your original idea.

2. While riding through the Everglades, you become separated from your friends when you decide to leave the boat and walk among the logs. Strangely, one of the logs appears to open its eyes. Finish the story.

LOCATION #79

EAST This way

1. The national motto of this country is, "Ddraig goch ddyry cychwyn." (The red dragon leads the way.)
2. Charles was investitured as Prince of this country in 1969.
3. This country has a tiny village with the longest place-name in the world:
    Llanfairpwllgwyngyllgogerychwyrndrobwllllandysilio!
4. The main character in Susan Cooper's "The Grey King" visits this country.

Name of country, province, or state: _____

Proof: _____

Creativity Across The Curriculum

1. People throughout the world enjoy an unusual old sport, known by the people of Country Number 79 as Gurning. The winner of this sport is the player who can make the funniest face. Experiment to come up with the face you would enter in a gurning contest. Check it out with a mirror. Invite a group of friends to join in the competition.

2. The thirteenth-century Conway Castle is an important landmark here. Design and draw a castle of the twenty-first century. List its most important features.

3. List the qualities that you feel a good leader should possess. Can you find any famous people from this country who possess(ed) many of these characteristics? Name them.

Name _____

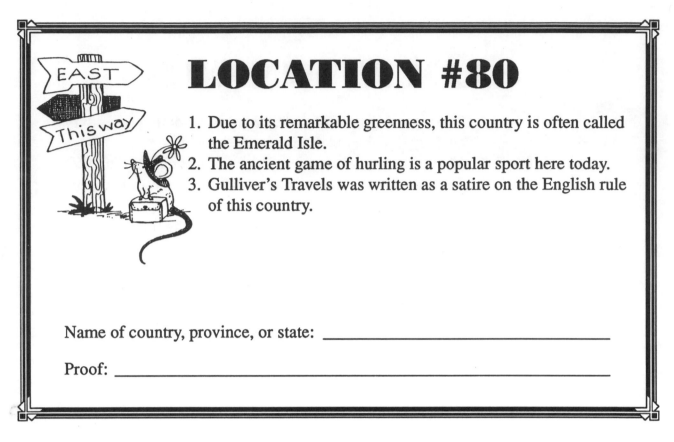

LOCATION #80

1. Due to its remarkable greenness, this country is often called the Emerald Isle.
2. The ancient game of hurling is a popular sport here today.
3. Gulliver's Travels was written as a satire on the English rule of this country.

Name of country, province, or state: _____

Proof: _____

Creativity Across The Curriculum

• Interview students to determine their favorite story characters from among the following categories:

    1. giants
    2. talking animals
    3. superheroes
    4. elves
    5. sports heroes
    6. real people

Graph your results.

• Create your own modern folktale using your town as the setting and using the language of today.

• Butter and marble are two important products of this country. Think of ten unusual uses for each one.

Name _____

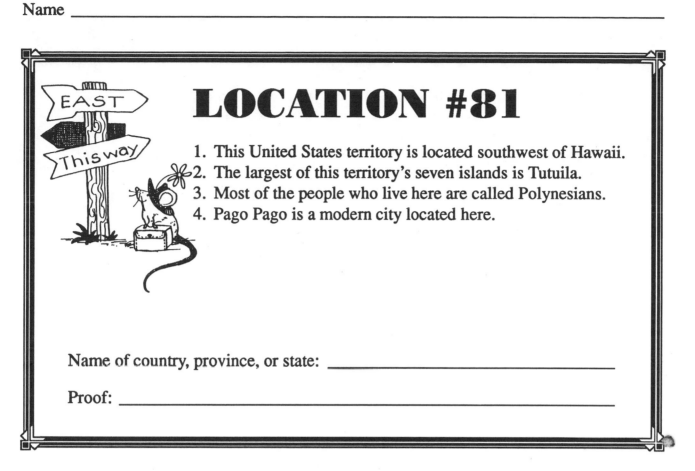

LOCATION #81

1. This United States territory is located southwest of Hawaii.
2. The largest of this territory's seven islands is Tutuila.
3. Most of the people who live here are called Polynesians.
4. Pago Pago is a modern city located here.

Name of country, province, or state: _____

Proof: _____

Creativity Across The Curriculum

1. Some of these islands are the remains of extinct volcanoes. What causes a volcano to erupt? Create a step-by-step cartoon of the process.

2. Yearly rainfall in this country is about 200 inches. How does this compare to the yearly rainfall in your state?

3. Taro is grown here. Draw it. Explain what it is and how it is used.

4. Design a game that could be played outdoors in a small amount of space on an island country. Draw the equipment you would need and write out the directions.

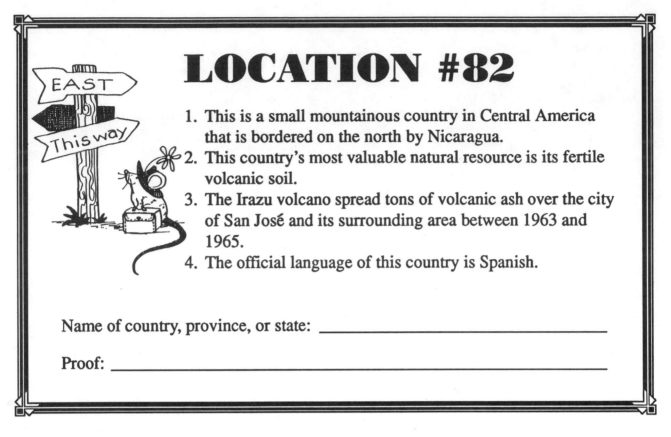

LOCATION #82

1. This is a small mountainous country in Central America that is bordered on the north by Nicaragua.
2. This country's most valuable natural resource is its fertile volcanic soil.
3. The Irazu volcano spread tons of volcanic ash over the city of San José and its surrounding area between 1963 and 1965.
4. The official language of this country is Spanish.

Name of country, province, or state: _____

Proof: _____

Creativity Across The Curriculum

1. Many people who live here in cities live in row houses. These are houses that look very similar and are attached to one another in a row. Using a pencil and ruler, create a block of eight row-houses—one of which would be for your family.

2. With a group of classmates, play a game of Spanish Bingo. Write a Spanish word from the list below in each square of each bingo card. The caller calls out the words in English. The players cover squares with the Spanish equivalents with their playing chips until one person covers a straight line of squares (vertically, horizontally, or diagonally) and calls out "Bingo."

Words:

uno = one	dos = two	tres = three
cuatro = four	cinco = five	seis = six
siete = seven	ocho = eight	nueve = nine
diez = ten	perrito = puppy	hola = hello
gracias = thanks	muy bien = very good	amigo = friend
por favor = please	chaqueta = coat	uva = grape
bonita = pretty	azucar = sugar	postres = desserts
tenedor = fork	pescado = fish	gaseosa = pop

B I N G O

		FREE		

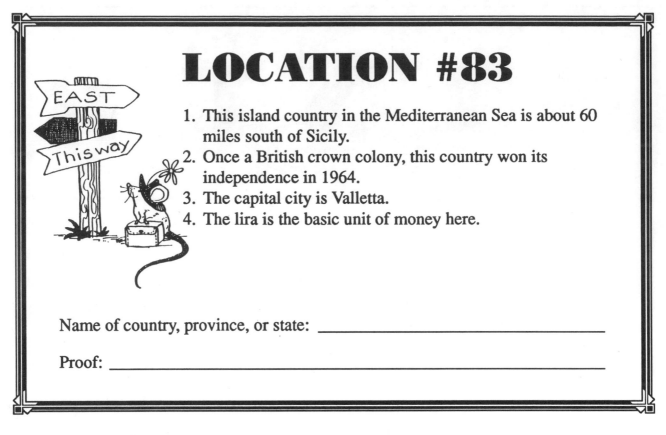

LOCATION #83

1. This island country in the Mediterranean Sea is about 60 miles south of Sicily.
2. Once a British crown colony, this country won its independence in 1964.
3. The capital city is Valletta.
4. The lira is the basic unit of money here.

Name of country, province, or state: _____

Proof: _____

Creativity Across The Curriculum

1. Make a bulletin board that highlights the most important island countries in the world. Entitle it, "Island Paradise." Make sure you include important information about each country's culture next to the name of each island. Develop a key to enable viewers to locate capital cities, rivers, mountain ranges, etc.

2. Customs in countries around the world have changed over the years. One way that change is easily noticed is observation of fashion trends. How have styles of clothing on the African continent changed in the last fifty years? Sketch a time line of men's and women's fashions.

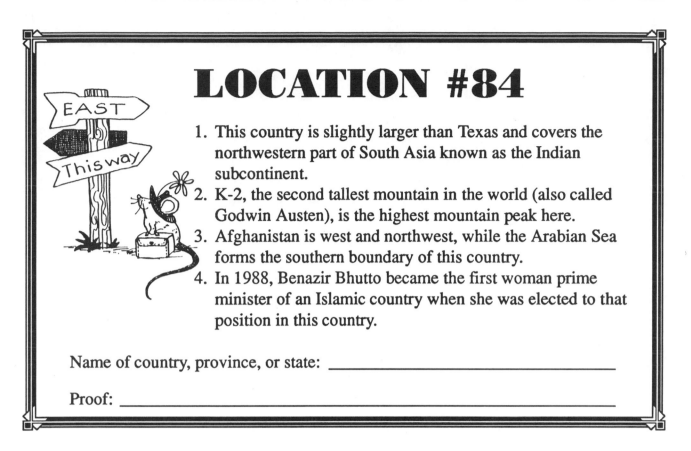

LOCATION #84

1. This country is slightly larger than Texas and covers the northwestern part of South Asia known as the Indian subcontinent.
2. K-2, the second tallest mountain in the world (also called Godwin Austen), is the highest mountain peak here.
3. Afghanistan is west and northwest, while the Arabian Sea forms the southern boundary of this country.
4. In 1988, Benazir Bhutto became the first woman prime minister of an Islamic country when she was elected to that position in this country.

Name of country, province, or state: _____

Proof: _____

Creativity Across The Curriculum

1. Camels, used in dry areas of this country for transportation and work, are moody animals. When they become angry, they whine and spit. Research the behavior of five animals of your choice to discover how they react when they become angry.

2. Because opium poppies grow easily with little water or care, the illegal production of the drugs opium and heroin has thrived in this country. We have had our "Just Say No" drug campaign in the U.S., with educational programs for children of all ages. What could be the slogan for this country? Create a drug awareness poster using this slogan.

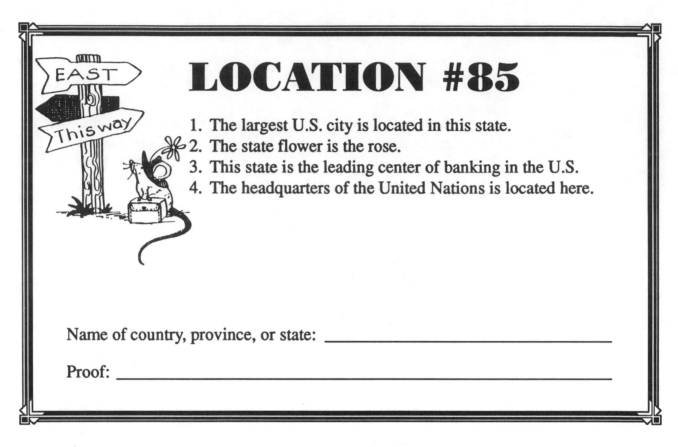

LOCATION #85

1. The largest U.S. city is located in this state.
2. The state flower is the rose.
3. This state is the leading center of banking in the U.S.
4. The headquarters of the United Nations is located here.

Name of country, province, or state: _____

Proof: _____

Creativity Across The Curriculum

1. Hirshfield, a caricaturist for the Times Newspaper, hid the word Nina in all of his drawings. Create a drawing of your own and hide your name in it.

2. If you were one of the original designers of the Statue of Liberty, what changes would you have made? Do some research to find out how we originally obtained the Statue of Liberty.

3. Complete "The People Explosion" activity.

The People Explosion

• Earth Day of 1990 made us all aware of the pollution and recycling problems of the earth. We now need to become increasingly aware of another problem of global concern—the increasing population of the earth. It stands to reason that an increase in population causes an increase in pollution on the planet. It is becoming difficult for the natural and economic resources of the earth to accommodate our "people explosion."

• Create a graph to show the enormity of the population problem by showing the following information.

U.S. Population in millions of people (rounded off)

1930—123 million (+ 16 percent)	1970—203 million (+ 13 percent)
1940—132 million (+ 7 percent)	1980—227 million (+12 percent)
1950—151 million (+14 percent)	1990—253 million (+11 percent)
1960—180 million (+ 19 percent)	2000—estimated at 280 million

What does this type of graph show you? What can you predict about the future? Do these figures tell scientists anything about future employment problems or other human needs?

• Create a second graph using the following information regarding changes in the population of the United States.

  1. A child is born every 8 seconds (10,800 per day)
  2. A person dies every 15 seconds (5,760 per day)
  3. 1 immigrant arrives every 33 seconds (2,618 per day)
  4. 1 emigrant leaves every 197 seconds (439 per day)

• What can we learn from this information? How will this affect such things as jobs, education, nutrition, pollution, and your life in general? Do some research to find out which areas of the world are growing the fastest. How does the United States compare to other countries with regard to its people explosion? How does Australia compare? What other information about Australia's population can you graph?

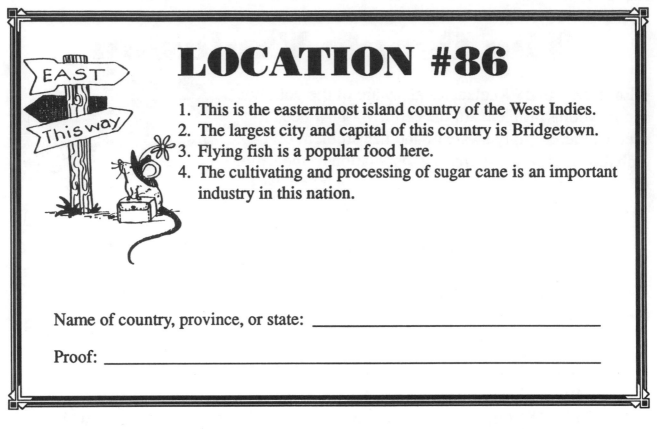

LOCATION #86

1. This is the easternmost island country of the West Indies.
2. The largest city and capital of this country is Bridgetown.
3. Flying fish is a popular food here.
4. The cultivating and processing of sugar cane is an important industry in this nation.

Name of country, province, or state: _____

Proof: _____

Creativity Across The Curriculum

1. Haiti, the Bahamas, Jamaica, and Puerto Rico are all island countries. After reading about each one, decide on which island you would most like to live. Design a postcard that you could send to family and friends that would tell about the country through pictures.

2. This island country has to import much of its food due to its large population and small land area. If you could import ten foods to add to your production of pork, sugar cane, milk, carrots, corn, and sweet potatoes, which food products would you import? List them in order of importance.

Flying Fish Wind Sock

Flying fish throw themselves from the water with their strong tails. Once in the air, their large fins act like wings, allowing them to fly up to 1,000 feet at a time. Create a windsock of bright paper to celebrate this wonderful species of fish. Follow these simple directions.

1 Begin with an 8½" x 11" piece of construction paper of your favorite color. Curl this into a cylinder and staple or tape.

2 Add 8 streamers of 1" x 11" contrasting-colored construction paper. Attach these to the base of the cylinder with staples, tape, or glue.

3 Create three brightly-colored fish on 8½" x 11" white construction paper. Make them fill the page and color them brightly. Cut them out and glue them around the cylinder. Suspend this from the ceiling with fish line.

STAPLER....

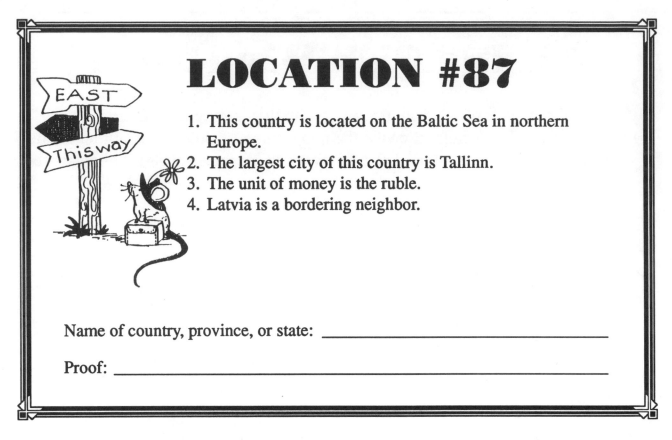

LOCATION #87

1. This country is located on the Baltic Sea in northern Europe.
2. The largest city of this country is Tallinn.
3. The unit of money is the ruble.
4. Latvia is a bordering neighbor.

Name of country, province, or state: _____

Proof: _____

Creativity Across The Curriculum

1. The swamps of this nation generate partly-decayed plant life called peat. This peat provides fuel for factories and power plants. The problem with peat is that it creates extensive air pollution due to the smoky fire it makes. What alternatives to peat could this nation use to provide for its fuel needs? Support your choices.

2. Create a crossword puzzle using words that promote a further understanding of this country.

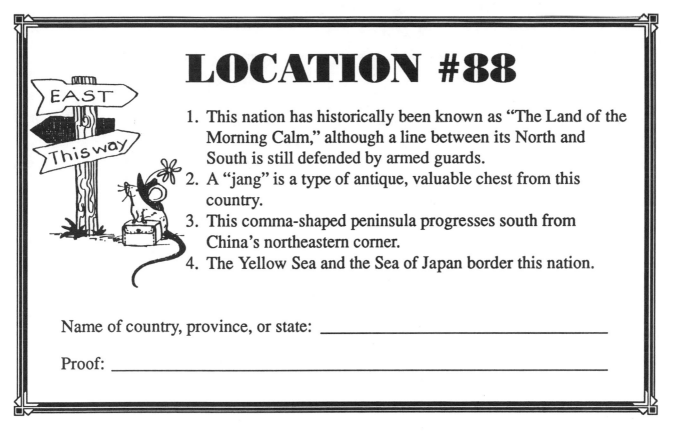

LOCATION #88

1. This nation has historically been known as "The Land of the Morning Calm," although a line between its North and South is still defended by armed guards.
2. A "jang" is a type of antique, valuable chest from this country.
3. This comma-shaped peninsula progresses south from China's northeastern corner.
4. The Yellow Sea and the Sea of Japan border this nation.

Name of country, province, or state: _____

Proof: _____

Creativity Across The Curriculum

1. The first encyclopedia made in the world was published here in the fifteenth century. This 112-volume work is found in the Library of Congress in Washington, D.C. If you were the head editor working on a new set of encyclopedias, what are the twenty-five most important topics you would want to include? List them in order of importance.

2. Dragons are a symbol of the ruler and protector from the east in Location Number 88 as well as in China. The dragon is often found on the eastern end of buildings as a symbol of protection. The Chinese dragons have five claws, and the dragons from Location Number 88 have four claws; thus, they can be distinguished from one another in art. Create an American dragon that could symbolize protection. Color it brightly, and describe its distinguishing features.

3. Make a list of activities that might occur here in the average day of an adult citizen.

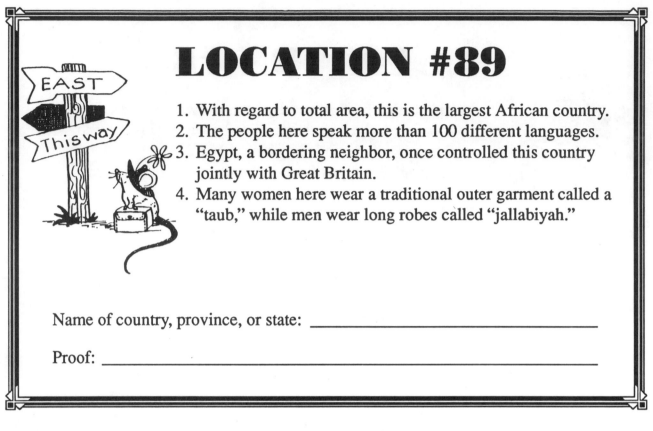

LOCATION #89

1. With regard to total area, this is the largest African country.
2. The people here speak more than 100 different languages.
3. Egypt, a bordering neighbor, once controlled this country jointly with Great Britain.
4. Many women here wear a traditional outer garment called a "taub," while men wear long robes called "jallabiyah."

Name of country, province, or state: _____

Proof: _____

Creativity Across The Curriculum

1. Less than one percent of the people of this country owned a television in 1989. Only one family out of fifteen had a radio, and five newspapers served the entire country. Think of creative ways to spread important news quickly to the people here. List five ideas.

2. What changes do you foresee happening in this country in the next twenty-five years? List ten.

3. In 1989 less than one percent of the population here owned a car. Design an inexpensive method of transportation for these farmers. Do not use gasoline for fuel.

Name _____

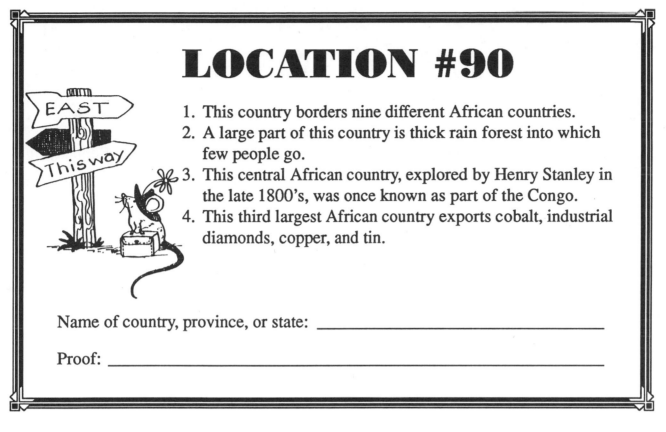

LOCATION #90

1. This country borders nine different African countries.
2. A large part of this country is thick rain forest into which few people go.
3. This central African country, explored by Henry Stanley in the late 1800's, was once known as part of the Congo.
4. This third largest African country exports cobalt, industrial diamonds, copper, and tin.

Name of country, province, or state: _____

Proof: _____

Creativity Across The Curriculum

1. The uranium for the first atomic bomb came from this country. Write a story about how our world would be different today if the atomic bomb had never been invented.

2. Part of this country is in the northern hemisphere and part is in the southern hemisphere. Explain how the two hemispheres are different.

3. The okapi can be found in this country. Create a postcard with an okapi on it.

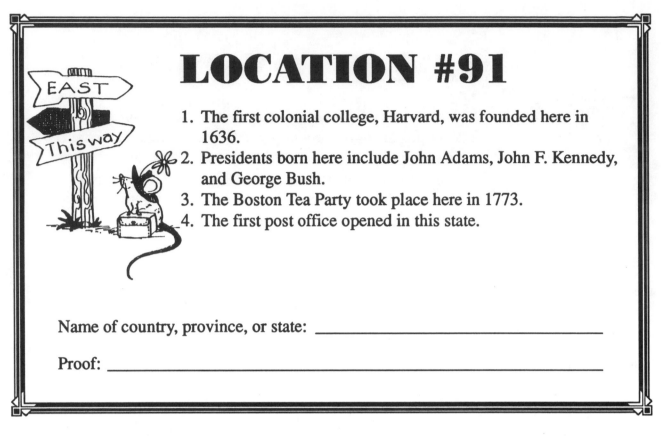

LOCATION #91

1. The first colonial college, Harvard, was founded here in 1636.
2. Presidents born here include John Adams, John F. Kennedy, and George Bush.
3. The Boston Tea Party took place here in 1773.
4. The first post office opened in this state.

Name of country, province, or state: _____

Proof: _____

Creativity Across The Curriculum

1. If the Pilgrims had never landed at Plymouth Rock in 1620, how would history have been changed?

2. Since the first post office opened in this state, design a stamp that honors its history.

3. If you could go back in time and spend a day with Paul Revere, what would you do for that day?

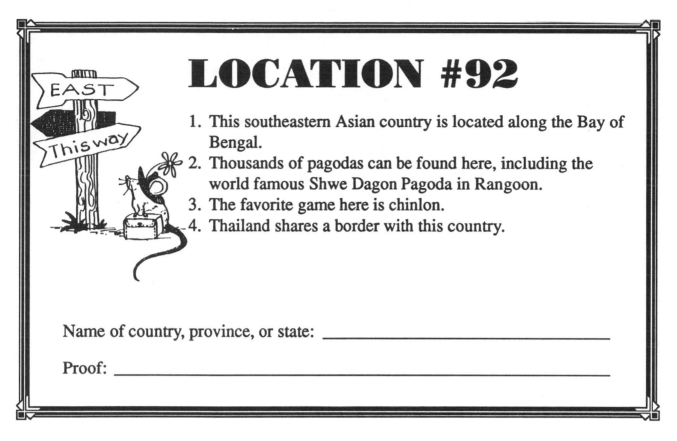

LOCATION #92

1. This southeastern Asian country is located along the Bay of Bengal.
2. Thousands of pagodas can be found here, including the world famous Shwe Dagon Pagoda in Rangoon.
3. The favorite game here is chinlon.
4. Thailand shares a border with this country.

Name of country, province, or state: _____

Proof: _____

Creativity Across The Curriculum

1. Most people in this country live in bamboo houses with thatch roofs. These houses are placed on poles above the ground to keep them safe from flooding and wild animals. Design another type of house that would work here. Take into consideration the natural resources of the country.

2. You have suddenly been transformed into a student from location Number 92. Write a letter to a pen pal from the U.S. telling him or her about your life. Be sure to include information about your siblings and parents, sports you like, your hobbies, foods you eat, your school, and your home.

Name _____

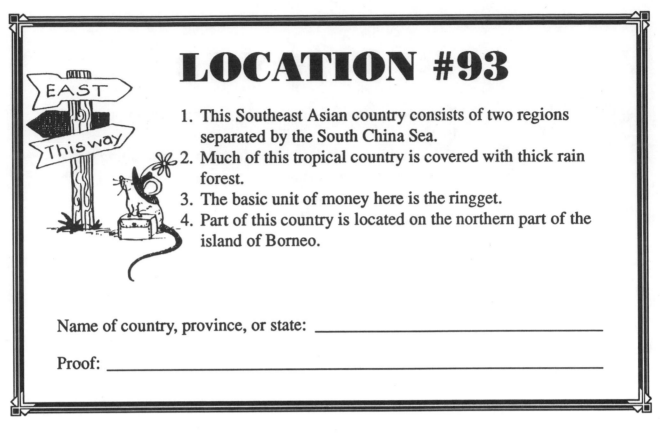

LOCATION #93

1. This Southeast Asian country consists of two regions separated by the South China Sea.
2. Much of this tropical country is covered with thick rain forest.
3. The basic unit of money here is the ringget.
4. Part of this country is located on the northern part of the island of Borneo.

Name of country, province, or state: _____

Proof: _____

Creativity Across The Curriculum

1. More than 500 kinds of birds are found in this country. Use your imagination to design a bird house that will house at least three different types of birds. Describe the birds that your house would attract.

2. The people of this country are very poor. What one new item would you invent to help these people? Use your imagination.

3. The unit of money here is the ringget. If you were going to design a new unit of money for the people of the U.S., what would it be? It has to be something completely different from anything that we already have. Sketch it, and tell what it is worth.

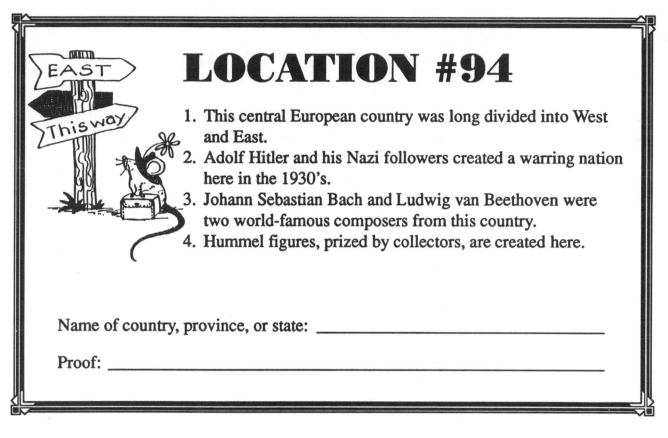

LOCATION #94

1. This central European country was long divided into West and East.
2. Adolf Hitler and his Nazi followers created a warring nation here in the 1930's.
3. Johann Sebastian Bach and Ludwig van Beethoven were two world-famous composers from this country.
4. Hummel figures, prized by collectors, are created here.

Name of country, province, or state: _____

Proof: _____

Creativity Across The Curriculum

1. This country has produced a great dichotomy of good and evil—great artists such as Beethoven and the evil political views of Adolph Hitler. What are our greatest "goods" and biggest "evils" in the U.S.? List five of each.

2. Compare the musical styles of Brahms, Mendelssohn, Schubert, and Schumann.

3. This country produces the Mercedes-Benz automobiles. Write an advertisement convincing American buyers to purchase a foreign-made car.

Name _____

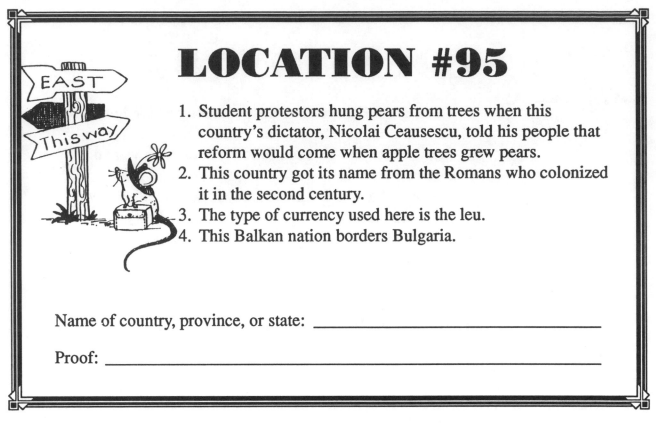

LOCATION #95

1. Student protestors hung pears from trees when this country's dictator, Nicolai Ceausescu, told his people that reform would come when apple trees grew pears.
2. This country got its name from the Romans who colonized it in the second century.
3. The type of currency used here is the leu.
4. This Balkan nation borders Bulgaria.

Name of country, province, or state: _____

Proof: _____

Creativity Across The Curriculum

1. What type of pet would you most likely have in this country? Think of ten unusual names for your pet. Which one would you choose? Why?

2. Plants of different types can often be grafted together to create new plants. One example might be— Pear + Apple = Pearple. Create two examples of your own. Tell as much as you can about each one. Then sketch them in their natural surroundings.

+   =

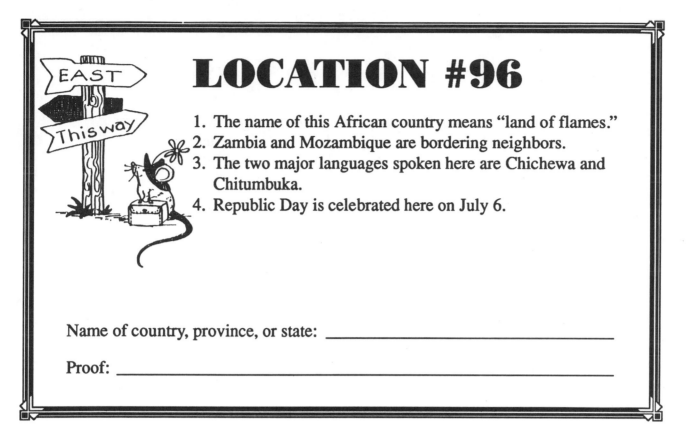

LOCATION #96

1. The name of this African country means "land of flames."
2. Zambia and Mozambique are bordering neighbors.
3. The two major languages spoken here are Chichewa and Chitumbuka.
4. Republic Day is celebrated here on July 6.

Name of country, province, or state: _____

Proof: _____

Creativity Across The Curriculum

1. The modern history of this African country began in 1859 when David Livingstone first came upon the third largest lake in Africa, located here. He called it Lake Nyasa. Livingstone was a missionary whose discovery led to further explorations of this area. Do some research to find out what you can about the life of David Livingstone. What were Livingstone's greatest contributions to the world?

2. In the history of all countries we find important people who have greatly contributed to our lives. Who do you know who might be likely to do each of the following things?

- Explore space
- Write a book
- Discover a cure for a disease
- Act in a movie
- Write a song
- Become president
- Become a famous athlete
- Create a new fashion
- Invent something new
- Become a lawyer

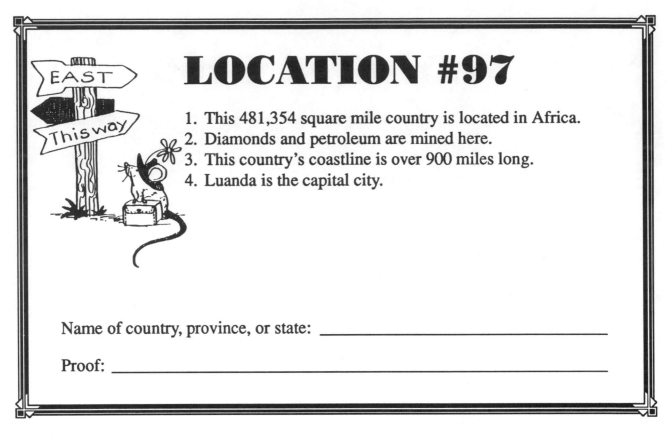

LOCATION #97

1. This 481,354 square mile country is located in Africa.
2. Diamonds and petroleum are mined here.
3. This country's coastline is over 900 miles long.
4. Luanda is the capital city.

Name of country, province, or state: _____

Proof: _____

Creativity Across The Curriculum

1. Cassava is one of the main crops produced here. Describe it and explain how it is used.

2. What do the black and red stripes on this country's flag symbolize? Create a new flag with a distinct meaning. Explain it.

3. This country, like many others around the globe, has been involved in civil wars. What are some reasons that civil wars begin? Why are they called "civil"? Write a prescription for preventing war.

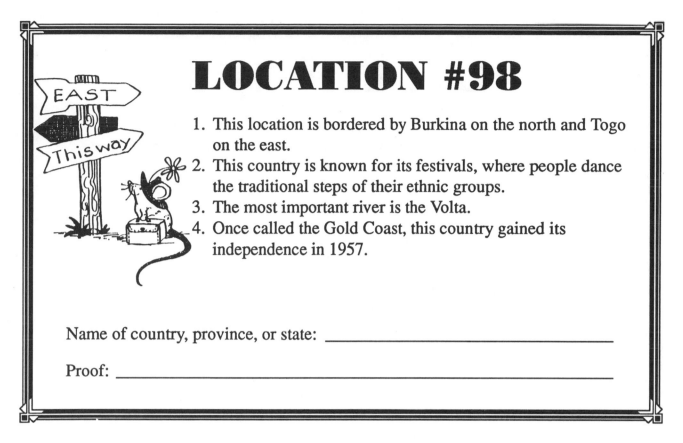

LOCATION #98

1. This location is bordered by Burkina on the north and Togo on the east.
2. This country is known for its festivals, where people dance the traditional steps of their ethnic groups.
3. The most important river is the Volta.
4. Once called the Gold Coast, this country gained its independence in 1957.

Name of country, province, or state: _____

Proof: _____

Creativity Across The Curriculum

1. Cacao is the chief cash crop in this country. How is cacao used? List ten uses.

2. The rain forest belt gives the people here the trees needed to support a large lumber industry. Write a letter to the leaders of this country convincing them to stop cutting down the rain forest.

3. Great Britain once ruled this country. List as many countries as you can that were once ruled by Great Britain.

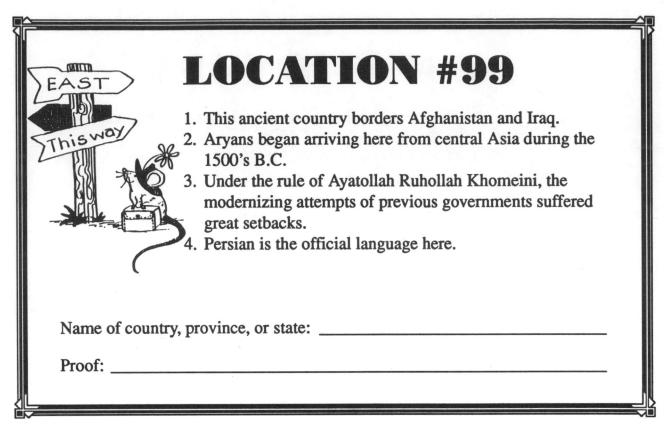

LOCATION #99

1. This ancient country borders Afghanistan and Iraq.
2. Aryans began arriving here from central Asia during the 1500's B.C.
3. Under the rule of Ayatollah Ruhollah Khomeini, the modernizing attempts of previous governments suffered great setbacks.
4. Persian is the official language here.

Name of country, province, or state: _____

Proof: _____

Creativity Across The Curriculum

1. The government of this country strongly recommends that women wear chadors—long body veils that are usually black and cover most of a woman's other clothes. How do you feel about the government telling people what to wear? List five reasons for your agreement or disagreement with the government's desire to do this.

2. A popular drink in this country is a thick yogurt drink called dough. Invent a new beverage and name it. Make a recipe card giving directions for making your new drink.

LOCATION #100

EAST
This way

1. Nelson Mandela, a well-known political freedom fighter from this country, was imprisoned for many years.
2. This country has three capital cities.
3. This country is the wealthiest country on the African continent.
4. Botswana is a bordering neighbor to the north.

Name of country, province, or state: _____

Proof: _____

Creativity Across The Curriculum

1. Aside from being a part of beautiful jewelry, how else can diamonds be of value to people? List five or more additional uses for diamonds.

2. Do some research to find news articles or book articles on apartheid. How has it changed the lives of the people of this country?

3. This country has three capitals. Choose one of these cities to become the one and only capital. Tell why you chose it.

Map Mania

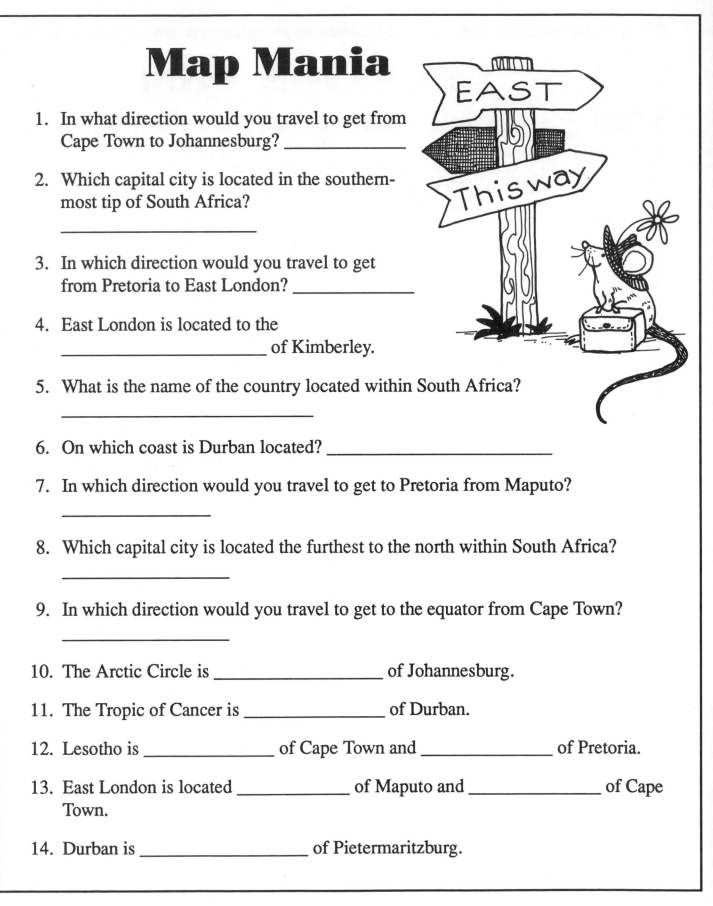

1. In what direction would you travel to get from Cape Town to Johannesburg? _____

2. Which capital city is located in the southern-most tip of South Africa?
   _____

3. In which direction would you travel to get from Pretoria to East London? _____

4. East London is located to the _____ of Kimberley.

5. What is the name of the country located within South Africa?
   _____

6. On which coast is Durban located? _____

7. In which direction would you travel to get to Pretoria from Maputo?
   _____

8. Which capital city is located the furthest to the north within South Africa?
   _____

9. In which direction would you travel to get to the equator from Cape Town?
   _____

10. The Arctic Circle is _____ of Johannesburg.

11. The Tropic of Cancer is _____ of Durban.

12. Lesotho is _____ of Cape Town and _____ of Pretoria.

13. East London is located _____ of Maputo and _____ of Cape Town.

14. Durban is _____ of Pietermaritzburg.

LOCATION #101

1. This state was the thirty-second state to become part of the United States.
2. Minneapolis and St. Paul are the Twin Cities located across the Mississippi River from each other.
3. This location is called the "Land of 10,000 Lakes."
4. The biggest man-made hole in the world is located here in the town of Hibbing.

Name of country, province, or state: _____

Proof: _____

Creativity Across The Curriculum

1. Paul Bunyan, the giant legendary figure of this area of the country, roamed the land with Babe the Blue Ox. Legend has it that wherever Babe walked, lakes were created. Create a legend of your own that will explain the creation of 10,000 lakes.

2. The lady's slipper is this state's state flower. Use your imagination to create a flower called man's biker boot. Tell where it would be found, which state's official flower it would be, and its size. Sketch and color it.

Name _____

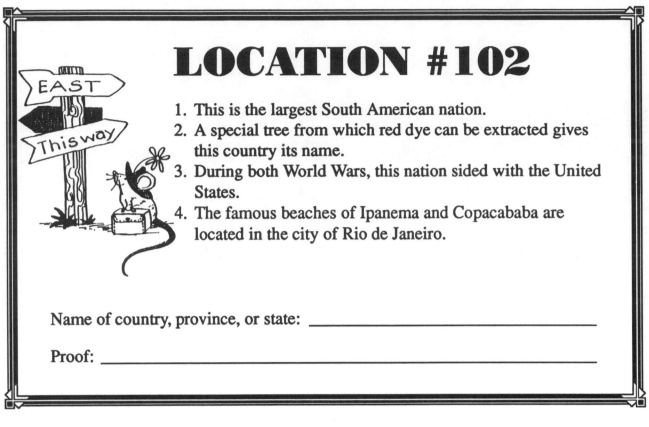

LOCATION #102

1. This is the largest South American nation.
2. A special tree from which red dye can be extracted gives this country its name.
3. During both World Wars, this nation sided with the United States.
4. The famous beaches of Ipanema and Copacababa are located in the city of Rio de Janeiro.

Name of country, province, or state: _____

Proof: _____

Creativity Across The Curriculum

1. Tropical rain forests are found in this country. Create five math problems that involve true rain forest facts. Each problem must require at least two separate operations.

2. Churrasco, a broiled, seasoned meat cut into chunks, is a native food of this country. From memory, write out the recipe of a favorite food of yours that is more likely to be found in the U.S. than in other parts of the world.

3. Of the crops grown in this country, which one would the people be most unable to get along without? Why?

4. Find something that happened 200 years ago in this country that is similar to something that is happening today somewhere else in the world.

Spotlight On Sports

Find the sports section of your local newspaper. Create five story headlines that you might find in newspapers in both the U.S. and Location 102.

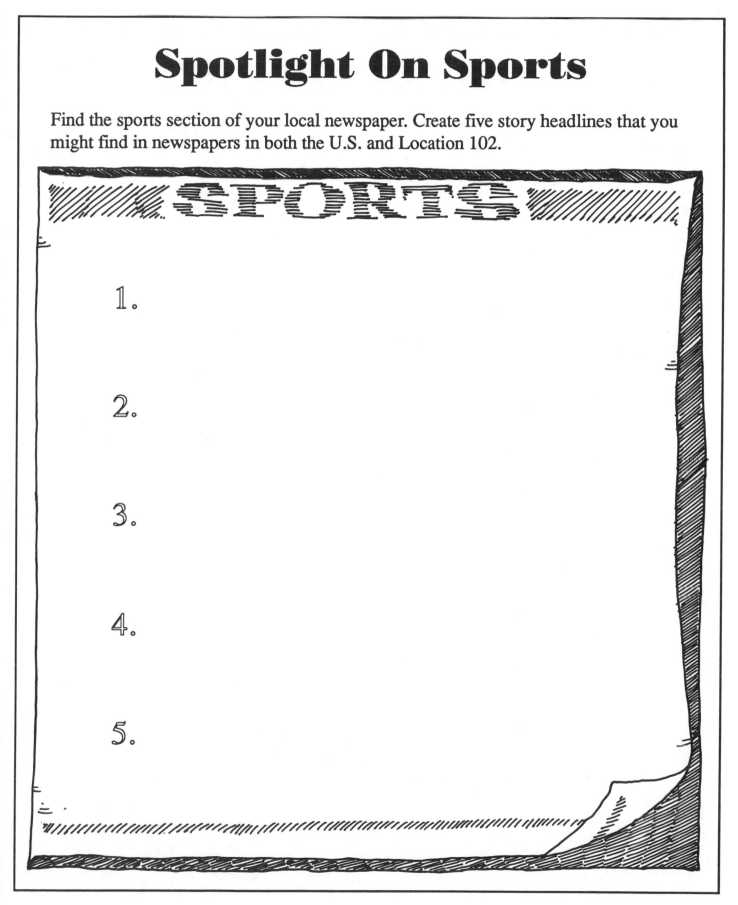

SPORTS

1.

2.

3.

4.

5.

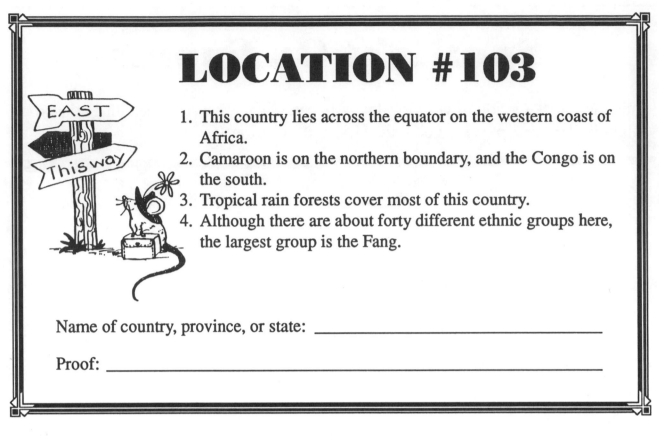

LOCATION #103

1. This country lies across the equator on the western coast of Africa.
2. Camaroon is on the northern boundary, and the Congo is on the south.
3. Tropical rain forests cover most of this country.
4. Although there are about forty different ethnic groups here, the largest group is the Fang.

Name of country, province, or state: _____

Proof: _____

Creativity Across The Curriculum

1. The wood of the okoume tree provides income for this country. Do research to find out how okoume wood is used.

2. The tropical rain forest is home to many plants, animals, and insects. List all of the flying insects that you can.

3. Brainstorm a list of movie titles that would best take place in and around a rain forest.

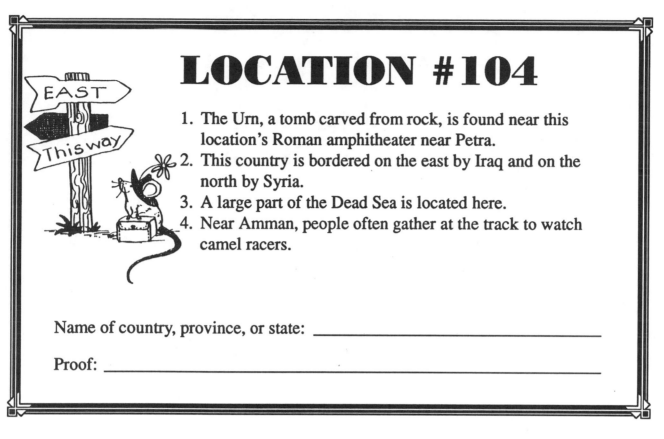

LOCATION #104

1. The Urn, a tomb carved from rock, is found near this location's Roman amphitheater near Petra.
2. This country is bordered on the east by Iraq and on the north by Syria.
3. A large part of the Dead Sea is located here.
4. Near Amman, people often gather at the track to watch camel racers.

Name of country, province, or state: _____

Proof: _____

Creativity Across The Curriculum

1. Visitors to this country can tour on camels by following ancient Roman caravan trails. Send a postcard to a friend of a member of your family describing what you have seen on your camel-back tour. (Hint: You will need to do some research on the tourist attractions found in this location.)

2. Some people prefer to live by the ocean and some prefer to live where it snows for a portion of the year. However, some people actually prefer to live in the desert. Create a poster advertising a home for sale in the desert. Be sure to highlight the advantages of the location.

Name _____

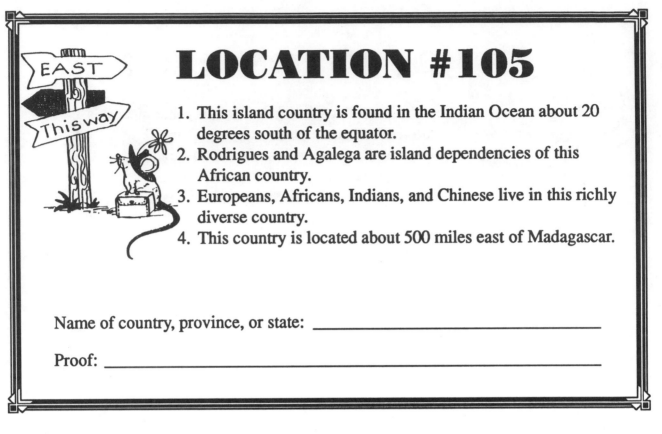

LOCATION #105

1. This island country is found in the Indian Ocean about 20 degrees south of the equator.
2. Rodrigues and Agalega are island dependencies of this African country.
3. Europeans, Africans, Indians, and Chinese live in this richly diverse country.
4. This country is located about 500 miles east of Madagascar.

Name of country, province, or state: _____

Proof: _____

Creativity Across The Curriculum

1. The diverse ethnic groups of this region provide a wide variety of customs. The Hindu-speaking Hindus make a yearly walk to a lake in the heart of the country to gather holy water. Many carry shrines and all are dressed in white clothing. The Tamil-speaking Hindus fulfill vows they have made to a variety of gods by walking on hot coals or climbing ladders of swords with the cutting edges turned up. Do some research to find some of the other cultural differences between Location Number 105 and your city, town, or state.

2. Many of the countries of Africa have been renamed several times. If you were to rename your state, what name would you choose? Give reasons for your choice.

3. Primarily Europeans, Africans, Indians, and Chinese live here. Which four ethnic groups provide the bulk of our population in the United States?

Name _____

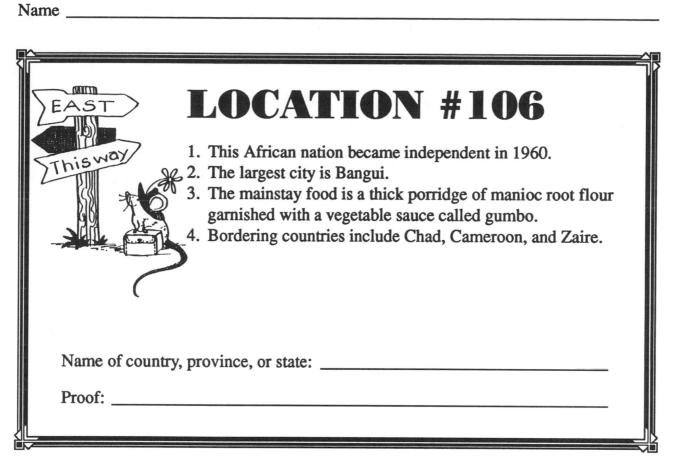

LOCATION #106

1. This African nation became independent in 1960.
2. The largest city is Bangui.
3. The mainstay food is a thick porridge of manioc root flour garnished with a vegetable sauce called gumbo.
4. Bordering countries include Chad, Cameroon, and Zaire.

Name of country, province, or state: _____

Proof: _____

Creativity Across The Curriculum

1. The southwestern part of this country has some tropical rain forest areas. Rain forests are part of the forest biome. It usually rains every day, and the temperature stays above 80 degrees fahrenheit. Chimpanzees, baboons, and some gorillas live here. Describe a typical day in one of these animal communities.

2. Huge termite mounds may be found in this country. Sketch a termite and list five facts that you could include in a book about this insect.

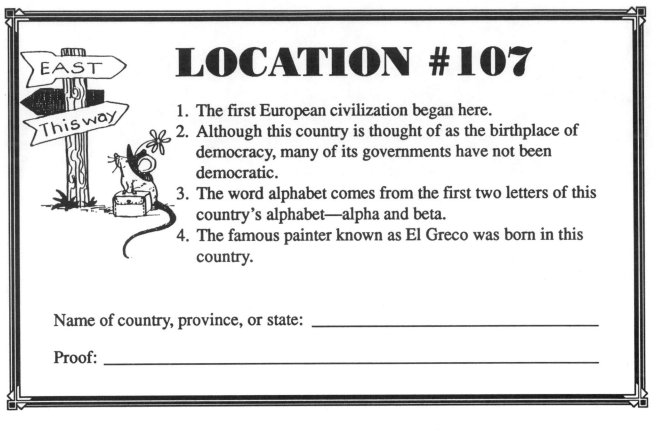

LOCATION #107

1. The first European civilization began here.
2. Although this country is thought of as the birthplace of democracy, many of its governments have not been democratic.
3. The word alphabet comes from the first two letters of this country's alphabet—alpha and beta.
4. The famous painter known as El Greco was born in this country.

Name of country, province, or state: _____

Proof: _____

Creativity Across The Curriculum

1. The Olympic Games were held every four years at Olympia to honor Zeus, the mythological king of the gods. Create a new olympic event and describe it in detail.

2. Read about several of this country's ancient mythological gods, such as Apollo and Athena. Create a myth, set in space in the year 2500, using some of these gods and goddesses as main characters.

3. Socrates, a famous philosopher from this country, taught by questioning his listeners. Support this technique as a good teaching strategy. What are other techniques that teachers use to help students learn?

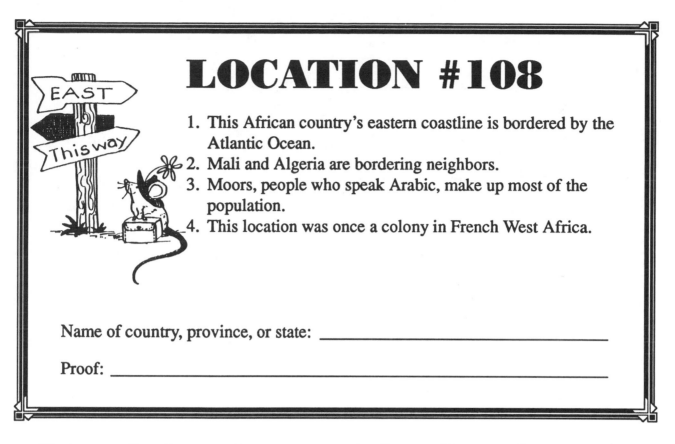

LOCATION #108

1. This African country's eastern coastline is bordered by the Atlantic Ocean.
2. Mali and Algeria are bordering neighbors.
3. Moors, people who speak Arabic, make up most of the population.
4. This location was once a colony in French West Africa.

Name of country, province, or state: _____

Proof: _____

Creativity Across The Curriculum

1. You have just traveled by camel across this country from east to west. Write an entry for your diary describing what you have seen.

2. Children in the United States begin early to think about a career. List the career options available to the people of this country. If you had to choose one, which career would you choose if you lived here? Why?

3. Design a product that would make life easier for the people who live here.

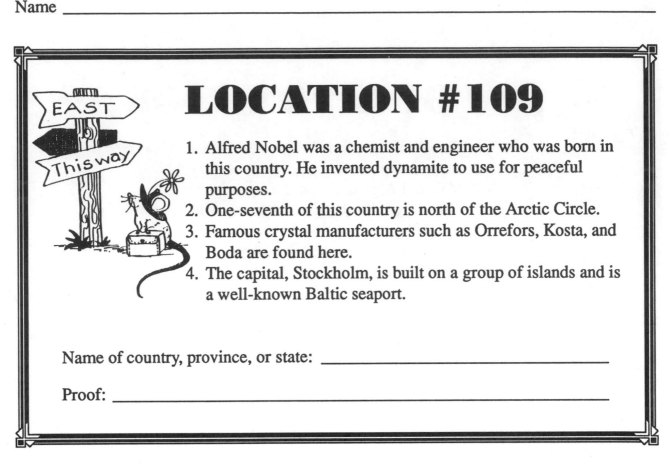

LOCATION #109

1. Alfred Nobel was a chemist and engineer who was born in this country. He invented dynamite to use for peaceful purposes.
2. One-seventh of this country is north of the Arctic Circle.
3. Famous crystal manufacturers such as Orrefors, Kosta, and Boda are found here.
4. The capital, Stockholm, is built on a group of islands and is a well-known Baltic seaport.

Name of country, province, or state: _____

Proof: _____

Creativity Across The Curriculum

1. The warship Vasa sank on the same day that it was launched in 1628. How did this ship compare to the Titanic? How were they different? The same?

2. The people of this country buy more magazines and newspapers than the people of any other country. If you wanted to begin publishing your own magazine, what would you name it? What types of stories would it contain? Write an example of one story.

What's Happening Here?

You are a reporter for one of the country's newspapers. Imagine you can look in a local newspaper. Create headlines to fill the following category chart.

CATEGORY CHART

FAMOUS PEOPLE	MONEY
ANIMAL STORY	BUSINESS
FOOD	TRANSPORTATION

Name _____

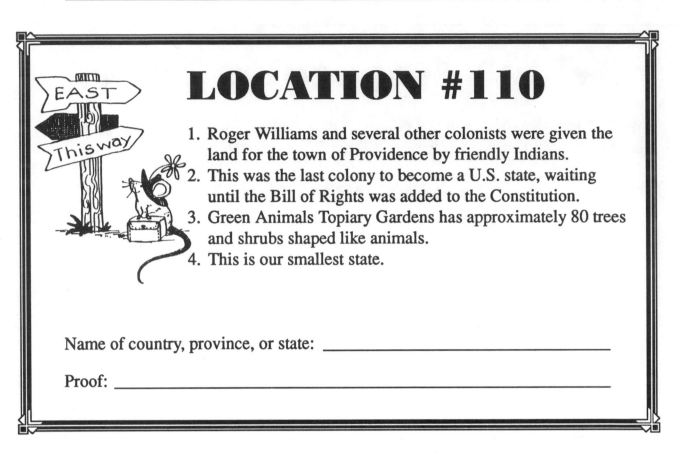

LOCATION #110

1. Roger Williams and several other colonists were given the land for the town of Providence by friendly Indians.
2. This was the last colony to become a U.S. state, waiting until the Bill of Rights was added to the Constitution.
3. Green Animals Topiary Gardens has approximately 80 trees and shrubs shaped like animals.
4. This is our smallest state.

Name of country, province, or state: _____

Proof: _____

Creativity Across The Curriculum

1. This state, like many others, is especially congested with traffic during "rush hour." It has more people per square mile than 48 of the states. Only New Jersey has more. To accommodate the growing population of workers (our world population doubled between 1930 and 1975), many countries have created various types of mass transit and high-speed transportation. Examples include Japan's "bullet train" and the Concorde jet. Design a transportation system that will assist city and suburban workers. Draw it and explain how it would operate.

2. This colony would not agree to become a state until the Bill of Rights became constitutional.
   a) Why do you think the Bill of Rights was so important to the members of this colony?
   b) How would our lives be changed today if the Bill of Rights did not exist?

Name _____

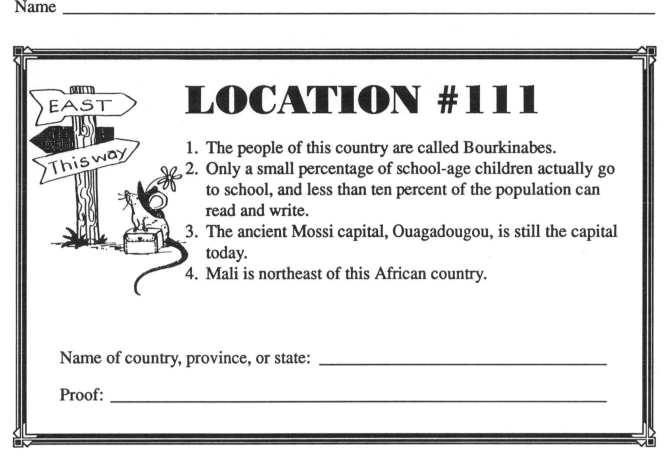

LOCATION #111

1. The people of this country are called Bourkinabes.
2. Only a small percentage of school-age children actually go to school, and less than ten percent of the population can read and write.
3. The ancient Mossi capital, Ouagadougou, is still the capital today.
4. Mali is northeast of this African country.

Name of country, province, or state: _____

Proof: _____

Creativity Across The Curriculum

1. This country has little in the way of art treasures. Design a piece of artwork that you would present to the country's leaders. Explain your choice.

2. Explain why this country would not be concerned about damage by permafrost.

3. Using clay, make a model of a home found in Country Number 111.

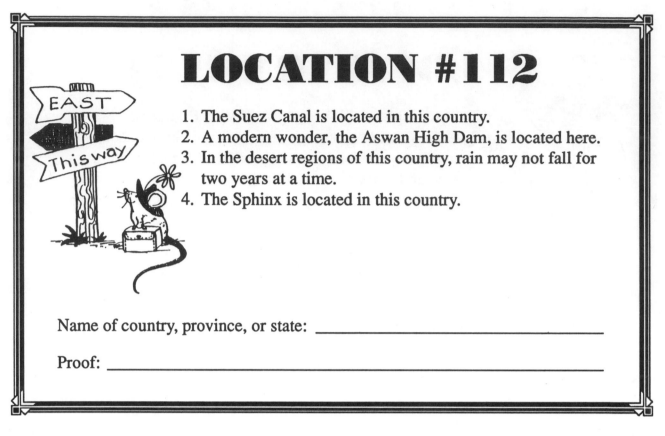

LOCATION #112

1. The Suez Canal is located in this country.
2. A modern wonder, the Aswan High Dam, is located here.
3. In the desert regions of this country, rain may not fall for two years at a time.
4. The Sphinx is located in this country.

Name of country, province, or state: _____

Proof: _____

Creativity Across The Curriculum

1. Report on an animal that is found only in Location Number 112.

2. A droodle is a drawing that doesn't make sense until the title is known. For example, this looks a beach ball, but it is really a giraffe walking past a porthole. Create your own droodle.

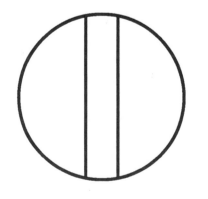

LOCATION #113

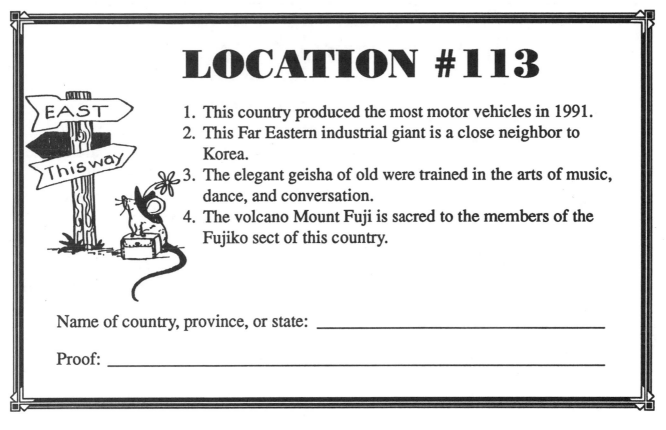

1. This country produced the most motor vehicles in 1991.
2. This Far Eastern industrial giant is a close neighbor to Korea.
3. The elegant geisha of old were trained in the arts of music, dance, and conversation.
4. The volcano Mount Fuji is sacred to the members of the Fujiko sect of this country.

Name of country, province, or state: _____

Proof: _____

Creativity Across The Curriculum

1. Suppose that when you wake up tomorrow morning, you are in Country Number 113 and are living with a family there. Describe your day from the time you awake until the time you go to bed.

2. Create something that reminds you of Country 113 from this incomplete line drawing.

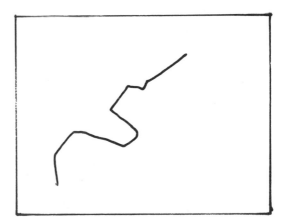

3. The writing from this country is very elegant and beautiful. Create a new mark for each letter of the alphabet. Use a small paint brush and dark paint to make your new alphabet.

4. Find out the type of poetry that is most often associated with Country Number 113. Try writing several of these poems. Illustrate your favorite.

Time And Magic

• Check the time right now. Write it down. Do some research to find out what time it is in Location Number 113. How long would it take you to fly from your present state to this country? What time would it be when you arrived there if you left in one hour?

• Read "Liang and the Magic Paintbrush" by Demi (Holt, Rhinehart, and Winston, New York, 1980). In this book Liang wants more than anything to paint. He mysteriously comes across a paintbrush that allows his artwork to come to life. Liang and his magic paintbrush outwit the greedy emperor in the story. Use your magic paintbrush to paint a picture that you wish would come to life. Do not allow yourself to be like the greedy emperor.

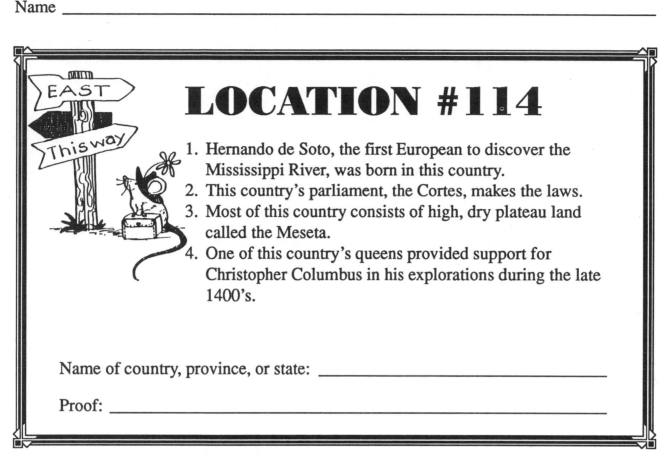

LOCATION #114

1. Hernando de Soto, the first European to discover the Mississippi River, was born in this country.
2. This country's parliament, the Cortes, makes the laws.
3. Most of this country consists of high, dry plateau land called the Meseta.
4. One of this country's queens provided support for Christopher Columbus in his explorations during the late 1400's.

Name of country, province, or state: _____

Proof: _____

Creativity Across The Curriculum

1. Complete this rhyme with four verses of your own:
   "My aunt came back from old Cape Town,
   And brought for me a pretty gown."
   "My aunt came back from old _____ ,
   And brought for me a _____."

2. Choose three distinctly different countries. Sketch the native clothing for all three. Ask a friend to try to discover the origin of each costume.

3. In the sea you will find plants that never, ever bloom. Name ten.

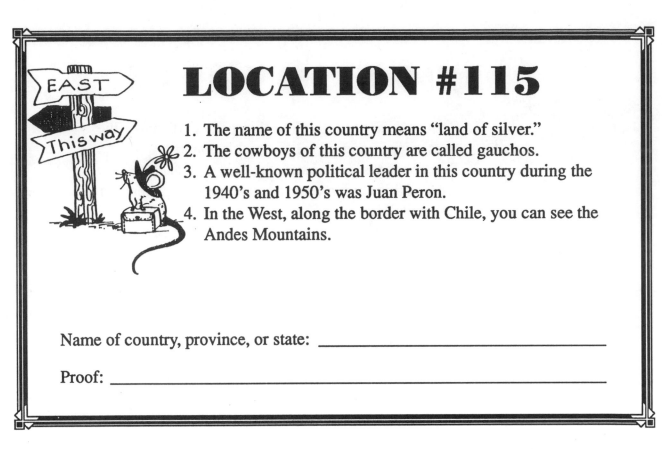

LOCATION #115

1. The name of this country means "land of silver."
2. The cowboys of this country are called gauchos.
3. A well-known political leader in this country during the 1940's and 1950's was Juan Peron.
4. In the West, along the border with Chile, you can see the Andes Mountains.

Name of country, province, or state: _____

Proof: _____

Creativity Across The Curriculum

1. The highest mountain here is Mount Aconcagua at 23,831 feet. Create a graph of the ten highest mountain ranges in the world. Scale your graph.

2. Using a piece of yarn, find the distance between the two largest South American cities. Prove your answer. (Hint: Use the scale of miles.)

3. How would the temperature in this country compare to the temperature in your city today? Create a line graph showing your city's average temperatures for a twelve-month period, along with Location Number 115's average temperatures.

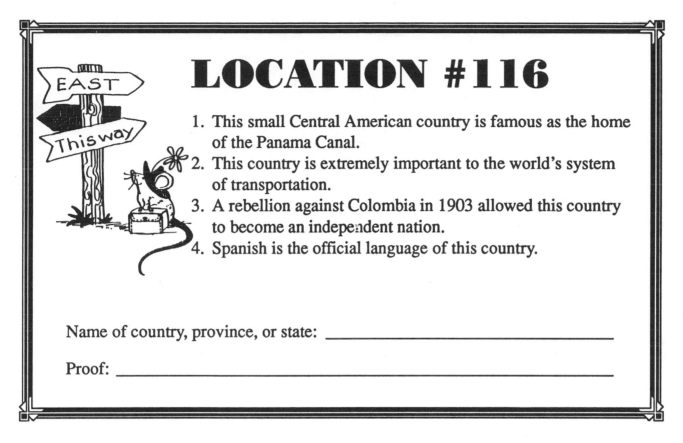

LOCATION #116

1. This small Central American country is famous as the home of the Panama Canal.
2. This country is extremely important to the world's system of transportation.
3. A rebellion against Colombia in 1903 allowed this country to become an independent nation.
4. Spanish is the official language of this country.

Name of country, province, or state: _____

Proof: _____

Creativity Across The Curriculum

1. The Panama Canal was built by the U.S. in 1914, connecting the Atlantic and Pacific oceans. By cutting through this canal, ships from many parts of the world avoid a long trip around the southern tip of South America. In 1999 the canal will become the official property of Location Number 116. Present a case for or against the giving up control of the canal by the U.S.

2. The San Blas Indian women are famous for their colorful costumes with elaborate embroidery. They also wear rings in their noses and large metal discs for earrings. Compare this type of Indian costume to that of a tribe of modern-day Native Americans.

3. Predict five cultural changes that this country will experience during the next twenty years.

Name _____

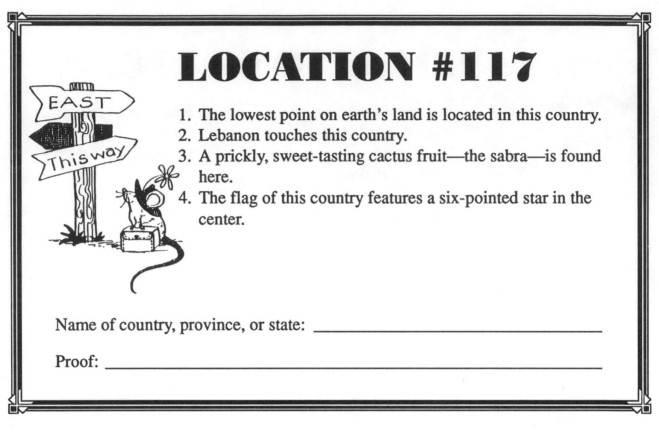

LOCATION #117

1. The lowest point on earth's land is located in this country.
2. Lebanon touches this country.
3. A prickly, sweet-tasting cactus fruit—the sabra—is found here.
4. The flag of this country features a six-pointed star in the center.

Name of country, province, or state: _____

Proof: _____

Creativity Across The Curriculum

1. Do some research to find the highest and lowest points on earth's land. What is the difference in feet between the highest and lowest points?

2. Draw a picture of an animal habitat that might be found at the lowest point on earth's land. Tell about it.

3. Research cactus plants from three continents.

LOCATION #118

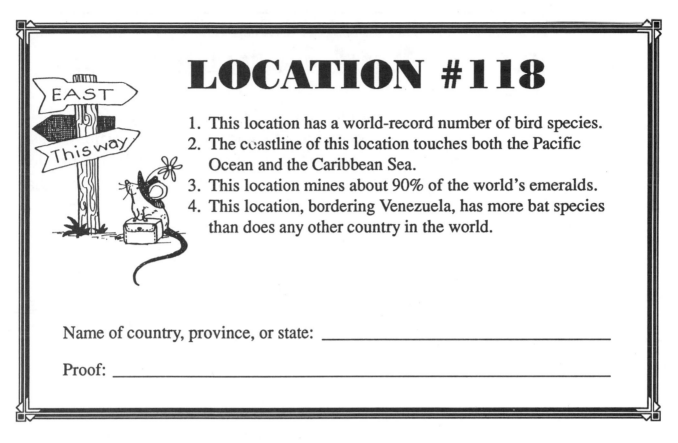

1. This location has a world-record number of bird species.
2. The coastline of this location touches both the Pacific Ocean and the Caribbean Sea.
3. This location mines about 90% of the world's emeralds.
4. This location, bordering Venezuela, has more bat species than does any other country in the world.

Name of country, province, or state: _____

Proof: _____

Creativity Across The Curriculum

1. Create a comparison chart of ten colorful birds found here. Compare by size, weight, location, color, diet, or other important feature.

2. Almost 95% of the world's emeralds come from this country. Research this most valuable of gems to discover how it is mined, how it is processed, and some old beliefs about its curative powers. Write down your findings.

Mystery And Comparison

• El Dorado, an Indian chief whose followers dusted him with gold on special occasions, was hunted by Spanish settlers who hoped to find his treasure. He was never found. Create a story that details the life and disappearance of El Dorado and finally solves the mystery.

• The bloodthirsty vampire bat is found in this country. Compare it to the fruit bat. How are they alike? Different? Sketch each one in its natural habitat. Which one do you find more interesting? Why?

Vampire Bat   **Fruit Bat**

Name _____

LOCATION #119

EAST
This way

1. This small country lies on the north coast of the island of Borneo.
2. The largest city is Bandar Seri Begawan.
3. The government here is headed by a monarch called a sultan.
4. This country borders the South China Sea on the north.

Name of country, province, or state: _____

Proof: _____

Creativity Across The Curriculum

1. Brainstorm a list of words to describe life on an island.

2. The discovery of oil has allowed this country's people to have a fairly comfortable life. What does the United States produce or create to allow many Americans to live comfortably?

3. What if:
  • Children in this country were not allowed to go to school?
  • You lived on this island?
  • The oceans were all so polluted that ocean life did not exist?
  • Every country in the world was required to speak only Spanish?

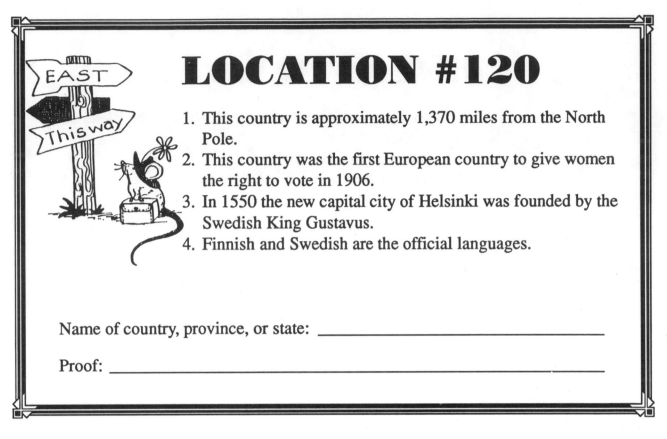

LOCATION #120

1. This country is approximately 1,370 miles from the North Pole.
2. This country was the first European country to give women the right to vote in 1906.
3. In 1550 the new capital city of Helsinki was founded by the Swedish King Gustavus.
4. Finnish and Swedish are the official languages.

Name of country, province, or state: _____

Proof: _____

Creativity Across The Curriculum

1. The forests of this country are its most important natural resource. The people take great care to protect their trees. In what ways can all people help to preserve the environment? List ten ways.

2. This country is the world's largest exporter of furs. The wearing of fur coats is a controversial issue in many places around the globe. List three reasons people should be allowed to wear and own fur coats and three reasons people should not be allowed to wear or own them.

3. Music is very important here. Finnlevy is an internationally famous producer of tapes and CD's. If you owned a company that produced musical tapes, who would you most want to record for you and why? Design a CD cover for this group.

LOCATION #121

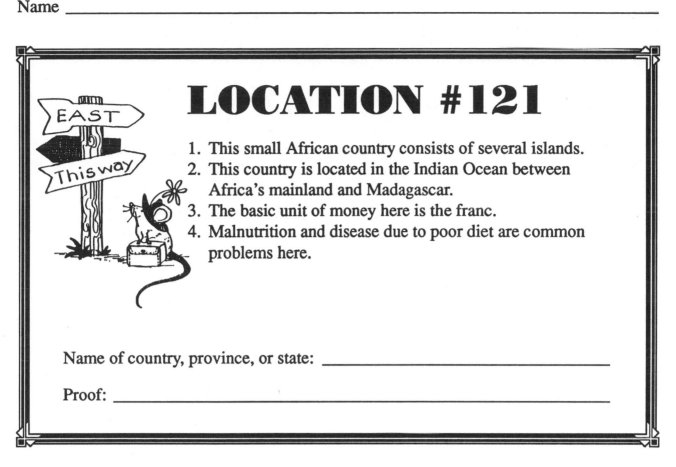

1. This small African country consists of several islands.
2. This country is located in the Indian Ocean between Africa's mainland and Madagascar.
3. The basic unit of money here is the franc.
4. Malnutrition and disease due to poor diet are common problems here.

Name of country, province, or state: _____

Proof: _____

Creativity Across The Curriculum

1. This country has a hot, rainy season from November to April. The people are often too poor to afford shoes. Think of ten ways that people could protect their feet if they didn't have shoes.

2. List ten cities east of Moroni and ten cities west of Moroni.

3. This country has very few modern conveniences. Without watches or clocks, how could people tell time?

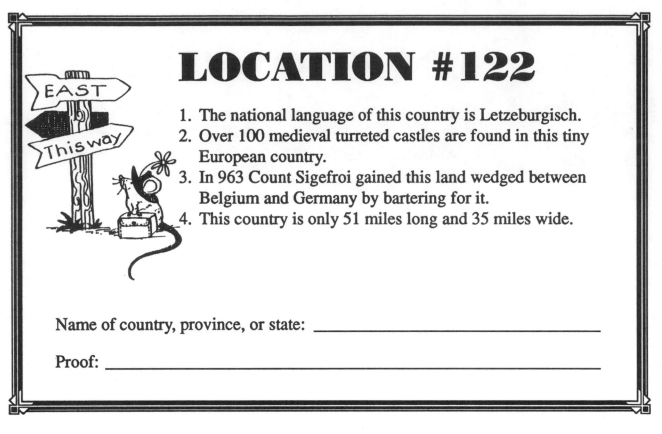

LOCATION #122

1. The national language of this country is Letzeburgisch.
2. Over 100 medieval turreted castles are found in this tiny European country.
3. In 963 Count Sigefroi gained this land wedged between Belgium and Germany by bartering for it.
4. This country is only 51 miles long and 35 miles wide.

Name of country, province, or state: _____

Proof: _____

Creativity Across The Curriculum

1. Create a table display to advertise traveling to Europe. Use books, magazines, newspapers, posters and memorabilia to attract the attention of others.

2. During The Battle of the Bulge in 1944, the Germans destroyed much of this country. Write a poem about war and a poem about peace. Cut construction paper background forms that symbolize these two concepts and mount your poems on them.

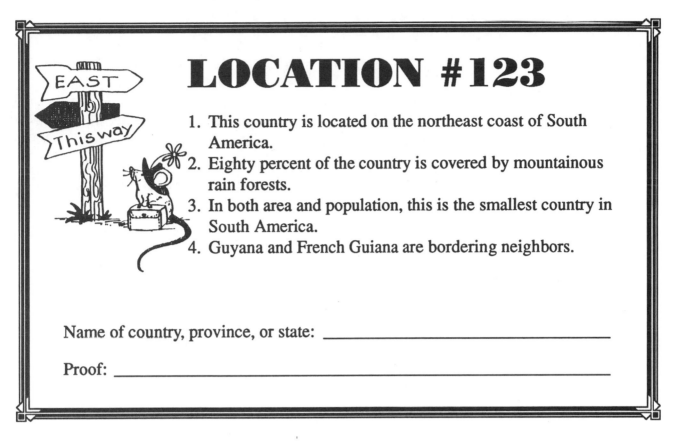

LOCATION #123

1. This country is located on the northeast coast of South America.
2. Eighty percent of the country is covered by mountainous rain forests.
3. In both area and population, this is the smallest country in South America.
4. Guyana and French Guiana are bordering neighbors.

Name of country, province, or state: _____

Proof: _____

Creativity Across The Curriculum

1. The river provides the main transportation route here. Design an effective method of water transportation that would be inexpensive to own and easy to maneuver around the rain forest waterways.

2. Write a story entitled, "The Day We Woke Up And Found Everyone's Skin Was The Same Color."

3. Create a picture using only one color.

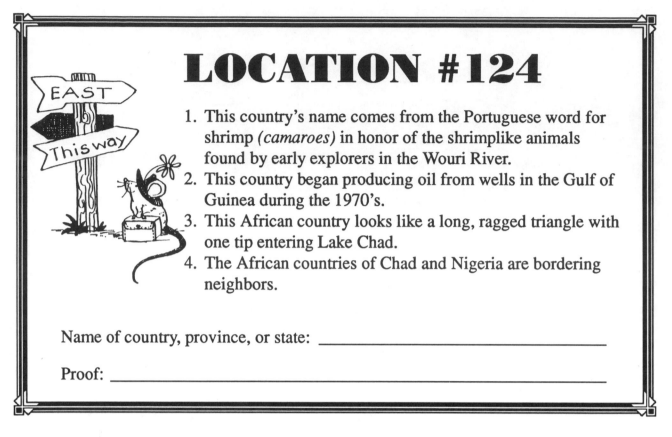

LOCATION #124

1. This country's name comes from the Portuguese word for shrimp *(camaroes)* in honor of the shrimplike animals found by early explorers in the Wouri River.
2. This country began producing oil from wells in the Gulf of Guinea during the 1970's.
3. This African country looks like a long, ragged triangle with one tip entering Lake Chad.
4. The African countries of Chad and Nigeria are bordering neighbors.

Name of country, province, or state: _____

Proof: _____

Creativity Across The Curriculum

1. A large percentage of the people here work on farms. Cassava is a crop they raise. Create an original cassava recipe.

2. Use three different sources of information to find out about this country's population, climate, chief products, education and flag. Create an informational card (bottom of page 161) for each topic, similar to the example at the top of that page. If the three sources agree in their information, place a star by each source. If two agree, place stars by the two that agree. If all three sources have different information, write your ideas as to why this occurs on the back of the card.

Verifying Sources

Example:

		CODE
Topic:		
History		
Source 1:	*Information:*	★
World Book C	Became unified in 1972	
Source 2:	*Information:*	★
Book of Knowledge C	Became independent in 1972	
Source 3:	*Information:*	★
Lands and Peoples Volume 1	New constitution in 1972	

		CODE
Topic:		
_____		
Source 1:	*Information:*	
_____	_____	
Source 2:	*Information:*	
_____	_____	
Source 3:	*Information:*	
_____	_____	

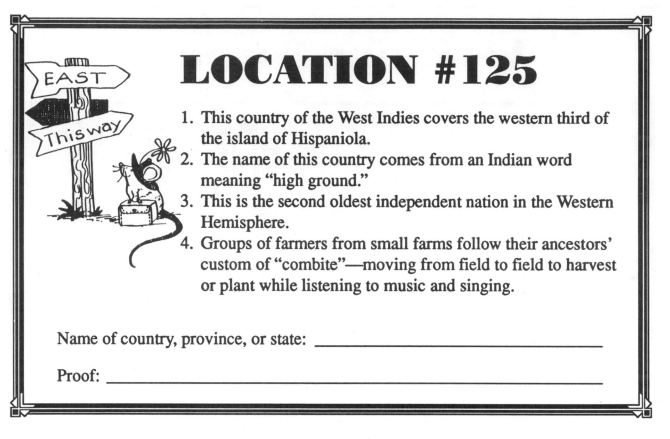

LOCATION #125

1. This country of the West Indies covers the western third of the island of Hispaniola.
2. The name of this country comes from an Indian word meaning "high ground."
3. This is the second oldest independent nation in the Western Hemisphere.
4. Groups of farmers from small farms follow their ancestors' custom of "combite"—moving from field to field to harvest or plant while listening to music and singing.

Name of country, province, or state: _____

Proof: _____

Creativity Across The Curriculum

1. This country exports sugar to the United States. Compare sugar to an artificial sweetener. Conduct a test. Prepare two packages of powdered drink mix in exactly the same way. The beverages should be identical except that you will use sugar in one and an artificial sweetener in the other. Take a taste-test survey with your class. See which beverage they prefer as they sample a small glass of each one. The samples should simply be marked A and B. Record the number of students who prefer A and the number of students who prefer B on a graph.

2. Design a revolutionary new type of packaging for sugar. Sketch and describe your ideas.

3. This small island country is quite crowded. People need to become more concerned about recycling their waste materials. Take something old and used and recycle it. Share your new product with the class.

©1993 by Incentive Publications, Inc., Nashville, TN.

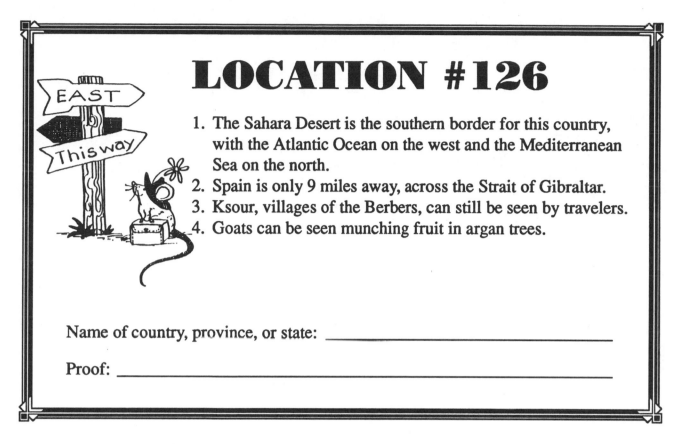

LOCATION #126

1. The Sahara Desert is the southern border for this country, with the Atlantic Ocean on the west and the Mediterranean Sea on the north.
2. Spain is only 9 miles away, across the Strait of Gibraltar.
3. Ksour, villages of the Berbers, can still be seen by travelers.
4. Goats can be seen munching fruit in argan trees.

Name of country, province, or state: _____

Proof: _____

Creativity Across The Curriculum

1. Most people in this country shop at the souk, a large market. Everything from food to goats to gold can be purchased at this open market. If you were a shopkeeper at an open market here, what would you sell? Explain your choice.

2. As do large cities everywhere, the large cities in this country have much crime. Design a machine that could tell whether or not a person had committed a crime. Who would you want to try it on first? Why?

3. Select two or three foods from this country that you have never eaten but would like to try. Draw and describe them.

Name _____

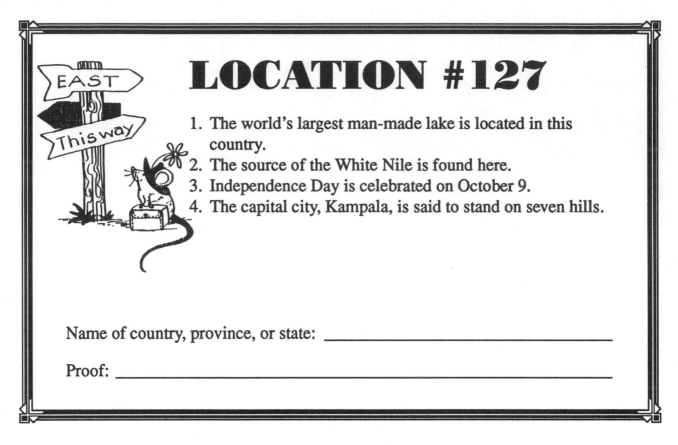

LOCATION #127

1. The world's largest man-made lake is located in this country.
2. The source of the White Nile is found here.
3. Independence Day is celebrated on October 9.
4. The capital city, Kampala, is said to stand on seven hills.

Name of country, province, or state: _____

Proof: _____

Creativity Across The Curriculum

1. Make a time line showing important historical dates for this country.

1400_____1450_____1500_____1550_____etc.

2. If, as adults, you and a group of seven friends and family members were to move to this country for one year, what types of jobs would you most likely find? How would your job choices be different if you stayed in the United States?

3. Copy an outline of this country onto typing paper. Create a face for it, and develop a story character. Give this character a name and develop an adventure.

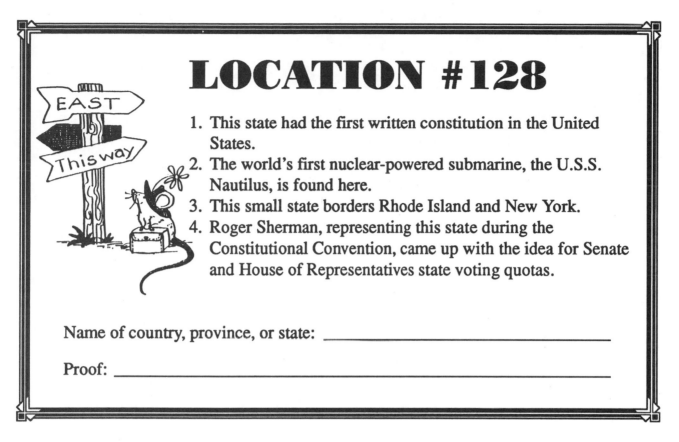

LOCATION #128

1. This state had the first written constitution in the United States.
2. The world's first nuclear-powered submarine, the U.S.S. Nautilus, is found here.
3. This small state borders Rhode Island and New York.
4. Roger Sherman, representing this state during the Constitutional Convention, came up with the idea for Senate and House of Representatives state voting quotas.

Name of country, province, or state: _____

Proof: _____

Creativity Across The Curriculum

1. In this state, Eli Whitney made guns with interchangeable parts. Ever since guns have become easier to make, more of them have been available to the general public. Laws have been written and changed regarding the ownership of guns by private citizens. If you were to make a law governing the rights of citizens to own guns, what would your law say? Write it out.

2. Listen to the song "Yankee Doodle." This song was popular with colonial soldiers during the Revolutionary War. Write another song, using the same tune, that you think would be appropriate for soldiers today.

3. Define treason. Why is this word important in American history?

Name _____

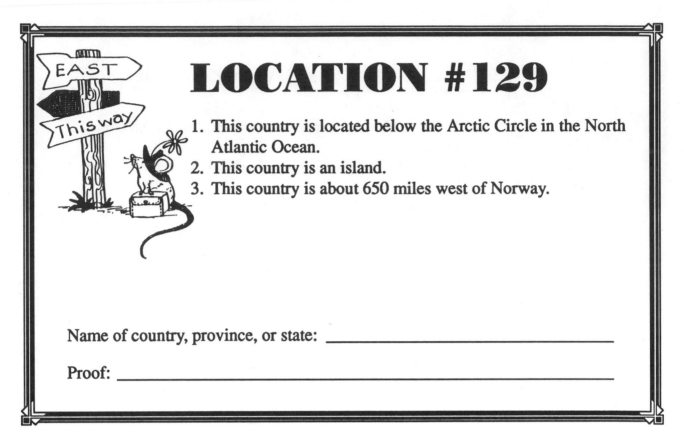

LOCATION #129

1. This country is located below the Arctic Circle in the North Atlantic Ocean.
2. This country is an island.
3. This country is about 650 miles west of Norway.

Name of country, province, or state: _____

Proof: _____

Creativity Across The Curriculum

1. The sale of fish and fish products provides the money needed to buy such things as appliances, food, and machinery. You are the owner of the largest container company in town. A group of fishermen have asked you to design a line of containers for their products. Draw a container for each of the following seafood needs:
   • shrimp
   • pickled herring
   • una
   • cod fillets

As part of the packaging design, create a logo for each container. The fishermen will call their company Safe Seafoods. Decide on shape, size, colors, logo, and written material for each container.

2. Disaster struck this location between 1400 and 1850 when the black death (bubonic plague) killed two-thirds of the island's people. Research the bubonic plague to see how the rest of the world was affected. Write a brief summary of your findings.

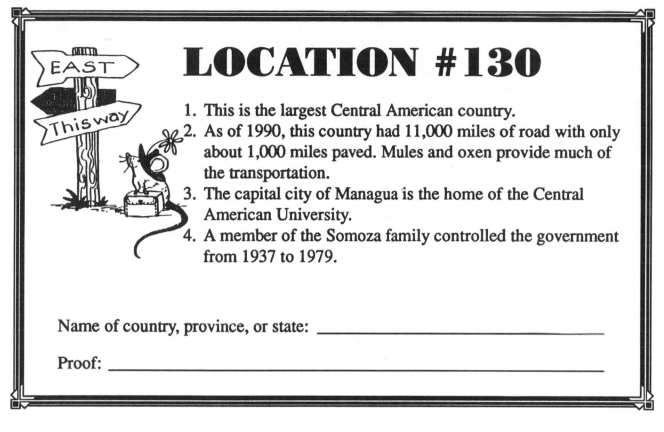

LOCATION #130

1. This is the largest Central American country.
2. As of 1990, this country had 11,000 miles of road with only about 1,000 miles paved. Mules and oxen provide much of the transportation.
3. The capital city of Managua is the home of the Central American University.
4. A member of the Somoza family controlled the government from 1937 to 1979.

Name of country, province, or state: _____

Proof: _____

Creativity Across The Curriculum

1. This country is slowly developing hydroelectric power. List five ways that you could cook your meals without electricity.

2. Corn, beans, and rice are the main crops raised here for food. Sketch three imaginative new ways to use each one.

3. Only the cities and towns of this country receive postal and telephone services. If you lived in a rural area here, how would your life be changed?

Name _____

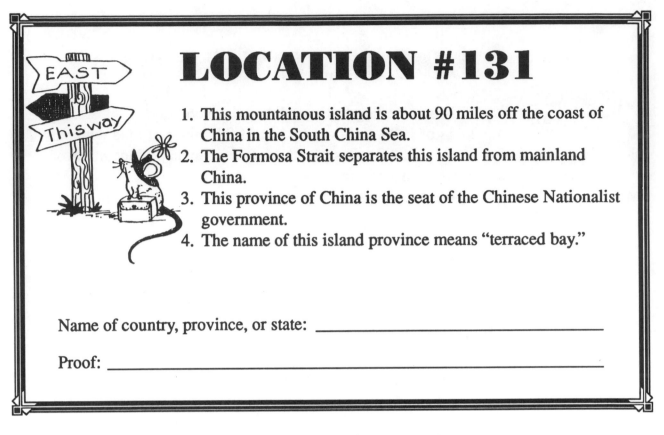

LOCATION #131

1. This mountainous island is about 90 miles off the coast of China in the South China Sea.
2. The Formosa Strait separates this island from mainland China.
3. This province of China is the seat of the Chinese Nationalist government.
4. The name of this island province means "terraced bay."

Name of country, province, or state: _____

Proof: _____

Creativity Across The Curriculum

1. This country is noted for manufacturing such things as calculators, televisions, and radios inexpensively. Check around your school and home for items made in Location #131. Make a list of your findings.

2. The first people to live in this country were the Aborigines. Do some research to find out about this group of people.

3. Do research to find out how this country is able to manufacture and sell electronic equipment for far less money than is the United States.

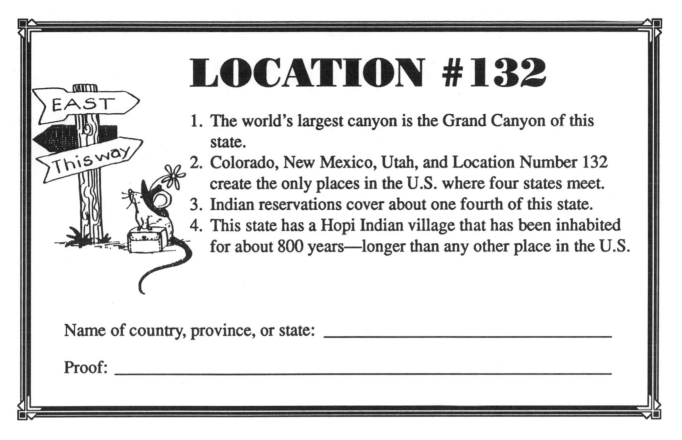

LOCATION #132

1. The world's largest canyon is the Grand Canyon of this state.
2. Colorado, New Mexico, Utah, and Location Number 132 create the only places in the U.S. where four states meet.
3. Indian reservations cover about one fourth of this state.
4. This state has a Hopi Indian village that has been inhabited for about 800 years—longer than any other place in the U.S.

Name of country, province, or state: _____

Proof: _____

Creativity Across The Curriculum

1. Create a legend recalling the formation of the Grand Canyon.

2. Compare the Grand Canyon to another important canyon.

3. You are a young Hopi Indian trying to recreate an historic tribal dwelling. Write out the directions for your construction and draw a sketch of the finished product.

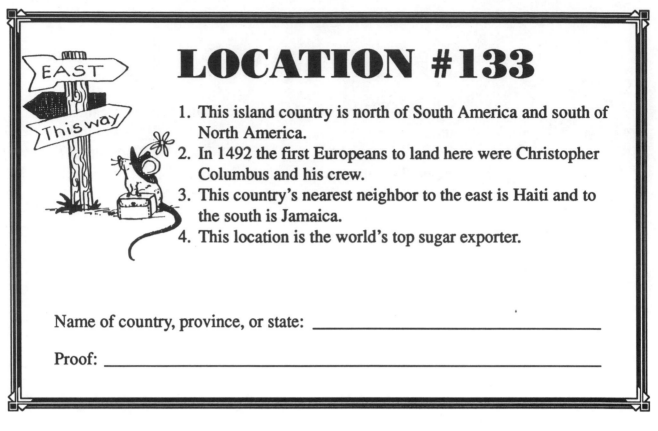

LOCATION #133

1. This island country is north of South America and south of North America.
2. In 1492 the first Europeans to land here were Christopher Columbus and his crew.
3. This country's nearest neighbor to the east is Haiti and to the south is Jamaica.
4. This location is the world's top sugar exporter.

Name of country, province, or state: _____

Proof: _____

Creativity Across The Curriculum

1. Create a survey that will allow you to find out what people enjoy or appreciate most about life in the United States.

2. Create a page for a class book entitled, "The Beauty of our World." Draw a beautiful landmark from somewhere in the world and create a brief paragraph of information to attach to the back.

3. Read "If I Were In Charge of the World," by Judith Viorst. Create a new verse of your own.

What Bugs You?

The people of Location 133 are bothered by some of the rules and regulations of their government. What "bugs" you about our United States government? Write down your top four ideas in the government bugs below.

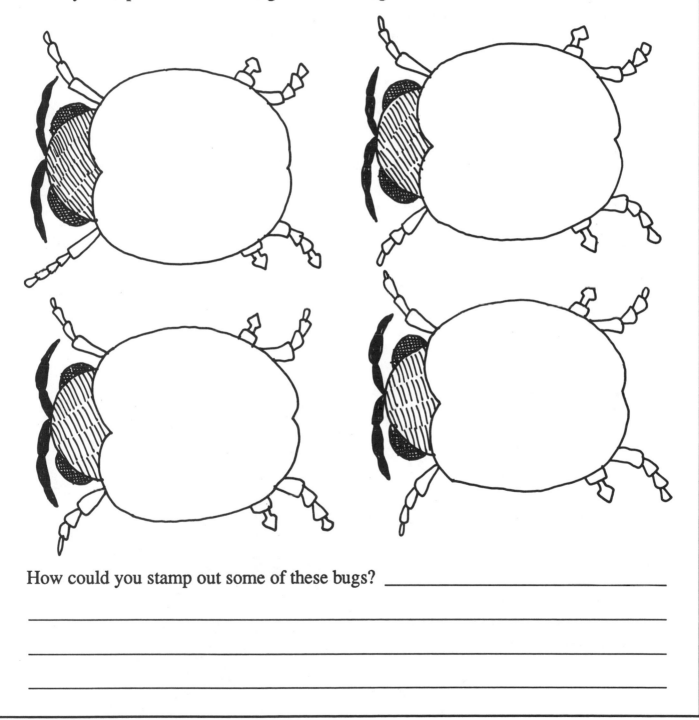

How could you stamp out some of these bugs? _____

_____

_____

_____

Name _____

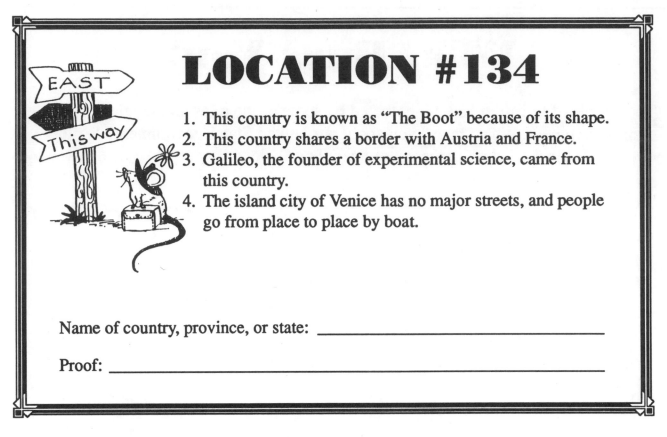

LOCATION #134

1. This country is known as "The Boot" because of its shape.
2. This country shares a border with Austria and France.
3. Galileo, the founder of experimental science, came from this country.
4. The island city of Venice has no major streets, and people go from place to place by boat.

Name of country, province, or state: _____

Proof: _____

Creativity Across The Curriculum

1. Famous operas such as Madame Butterfly and La Bohème are very much a part of this country's history. Brainstorm a list of ten words that you would use to describe opera to a person who had never heard one.

2. List some famous art works created by Michelangelo. Which is your favorite? Why?

3. Whom do you consider to be the five most important persons in this country's history? Why?

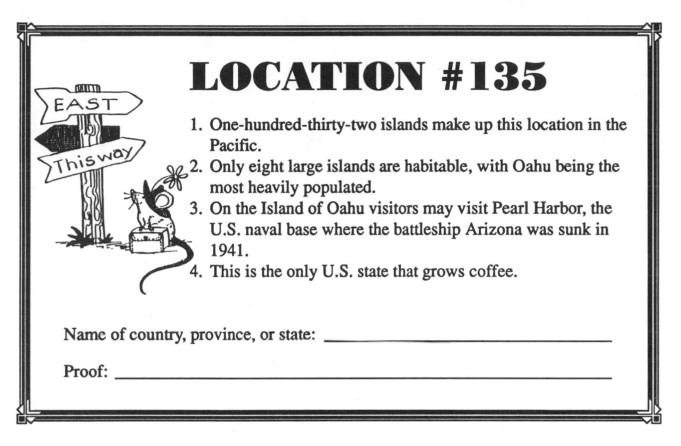

LOCATION #135

1. One-hundred-thirty-two islands make up this location in the Pacific.
2. Only eight large islands are habitable, with Oahu being the most heavily populated.
3. On the Island of Oahu visitors may visit Pearl Harbor, the U.S. naval base where the battleship Arizona was sunk in 1941.
4. This is the only U.S. state that grows coffee.

Name of country, province, or state: _____

Proof: _____

Creativity Across The Curriculum

1. Write an employment ad for a sugarcane field worker, encouraging him or her to come from the United States mainland to this island state.

2. Polynesian dancers tell stories with their hands as they dance. Learn a short story and devise a way to tell it with your hands.

3. Do some research to find out the facts behind the sinking of the battleship Arizona. How did this single act affect the history of the world?

Name _____

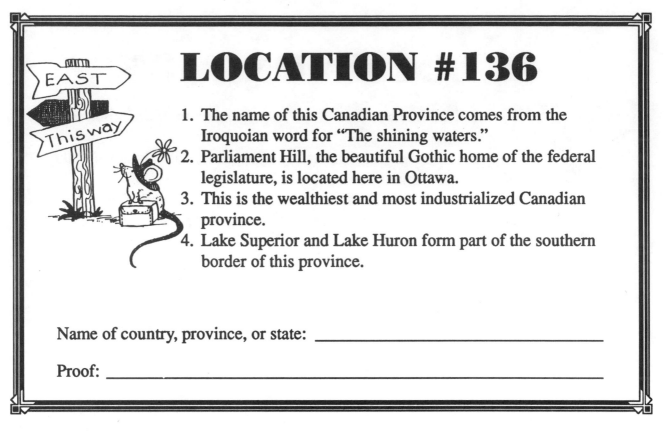

LOCATION #136

1. The name of this Canadian Province comes from the Iroquoian word for "The shining waters."
2. Parliament Hill, the beautiful Gothic home of the federal legislature, is located here in Ottawa.
3. This is the wealthiest and most industrialized Canadian province.
4. Lake Superior and Lake Huron form part of the southern border of this province.

Name of country, province, or state: _____

Proof: _____

Creativity Across The Curriculum

1. If you went to the library to find a book about this Canadian province, the Dewey Decimal System would allow you to easily find a book on Canada by looking for its call number. Design a new way to organize books in a library. Write out simple step-by-step instructions for your new procedure that could be followed by any librarian starting a new library.

2. If you were visiting this country for five days, how would you spend your time? Record the time you would spend on each activity. Be sure to include sleeping and eating.

3. Explain, by writing a letter to your principal, how you feel about children missing days of school to go on a family vacation.

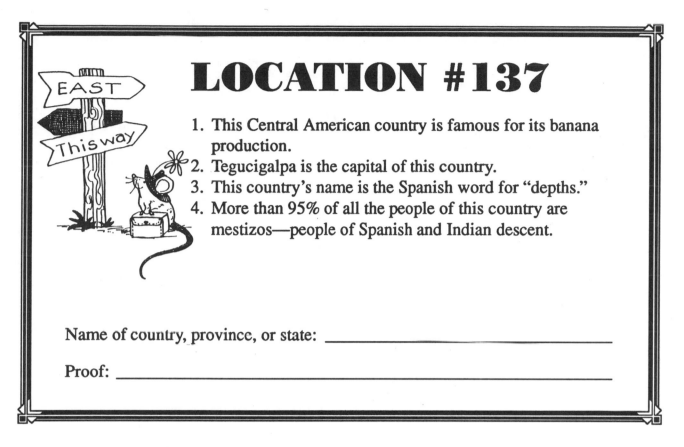

LOCATION #137

1. This Central American country is famous for its banana production.
2. Tegucigalpa is the capital of this country.
3. This country's name is the Spanish word for "depths."
4. More than 95% of all the people of this country are mestizos—people of Spanish and Indian descent.

Name of country, province, or state: _____

Proof: _____

Creativity Across The Curriculum

1. Create a triplet (three-line poem) about Location Number 137.
   Example:
   When Christopher Columbus
     landed here,
   During Isabella's reign,
   He claimed this land for Spain.

2. Cut words from magazines and newspapers to create a picture of this country. Put them together in an idea collage.

3. Many of the people living in this country are very poor. Draw a picture that gives you the feelings and emotions of the word "poor."

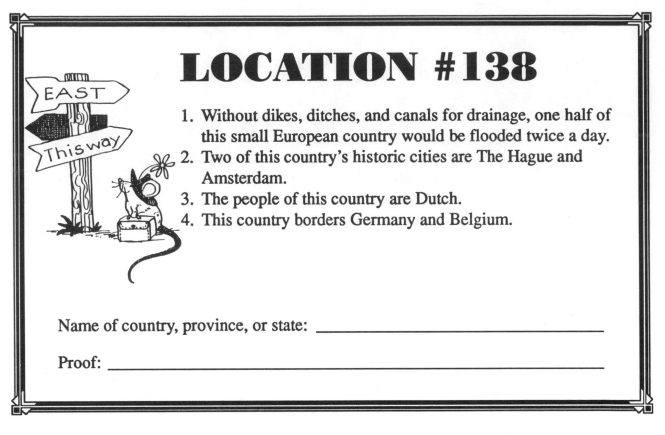

LOCATION #138

1. Without dikes, ditches, and canals for drainage, one half of this small European country would be flooded twice a day.
2. Two of this country's historic cities are The Hague and Amsterdam.
3. The people of this country are Dutch.
4. This country borders Germany and Belgium.

Name of country, province, or state: _____

Proof: _____

Creativity Across The Curriculum

1. Due to its large population, this country's houses are often only one room wide and several stories tall. Design the floorplan for your house if it can be only one room wide and five floors tall.

2. Compare the flag of Country Number 138 with the flag of Luxembourg. How are they different?

3. Windmills were once important to this country. List five other ways to produce electricity.

Name _____

LOCATION #139

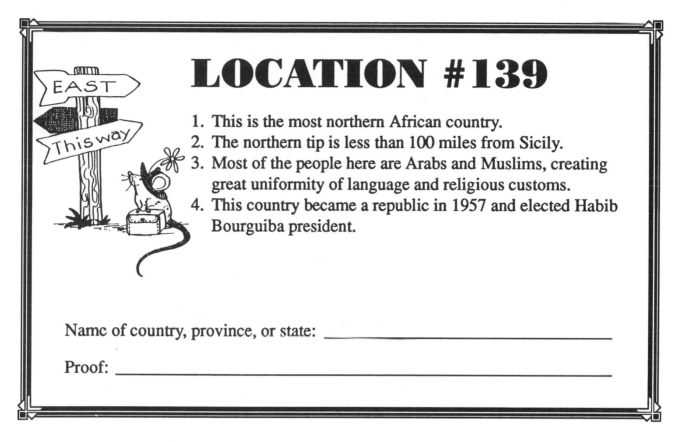

1. This is the most northern African country.
2. The northern tip is less than 100 miles from Sicily.
3. Most of the people here are Arabs and Muslims, creating great uniformity of language and religious customs.
4. This country became a republic in 1957 and elected Habib Bourguiba president.

Name of country, province, or state: _____

Proof: _____

Creativity Across The Curriculum

1. Compare the weather of this most northern African country to that of the most southern African country. Which weather would you prefer? Why?

2. In rural parts of this country you can still find traditional Arab clothing, such as a turban and a long, loose gown. In America, clothing is often viewed as a symbol of wealth or personal taste.
   - What can you tell about a person based on his or her clothes?
   - What are your favorite kinds of clothes? Why?

3. Create a t-shirt logo that you feel would be a fashion statement about who you are.

Name _____

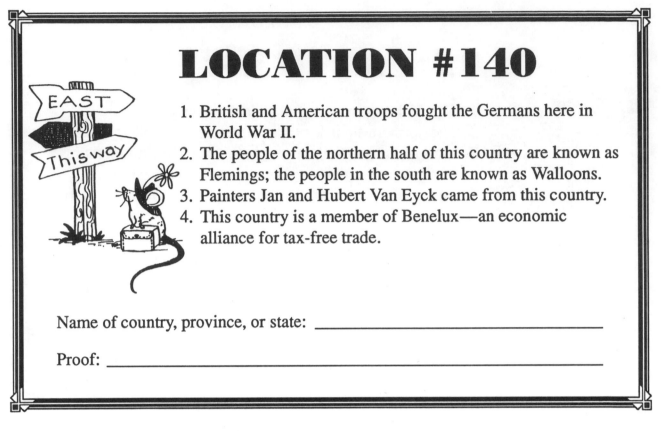

LOCATION #140

1. British and American troops fought the Germans here in World War II.
2. The people of the northern half of this country are known as Flemings; the people in the south are known as Walloons.
3. Painters Jan and Hubert Van Eyck came from this country.
4. This country is a member of Benelux—an economic alliance for tax-free trade.

Name of country, province, or state: _____

Proof: _____

Creativity Across The Curriculum

1. Adolphe Sax is best known for inventing the saxophone. What other instruments did he invent? Which one is your favorite? Why?

2. Draw a map that would show an explorer how to get to Location Number 140 from your home.

3. A plane can fly across this country in twenty minutes. How long would it take to fly from the east coast to the west coast in the United States? How does the size of Country Number 140 compare to the size of the United States?

LOCATION #141

1. This African country was once known as Abyssinia.
2. Sudan is to the west, and the Red Sea is on the northeast.
3. The symbol of this country is the lion.
4. The modern capital is Addis Ababa.

Name of country, province, or state: _____

Proof: _____

Creativity Across The Curriculum

• Disease is a great problem in this country. The average life span of people in this country is about 40 years. Compare this to the average life span of people in ten other countries of the world. Rank order your list from longest to shortest life expectancy.

Longest:

_____   _____

_____   _____

_____   _____

_____   Shortest:

_____   _____

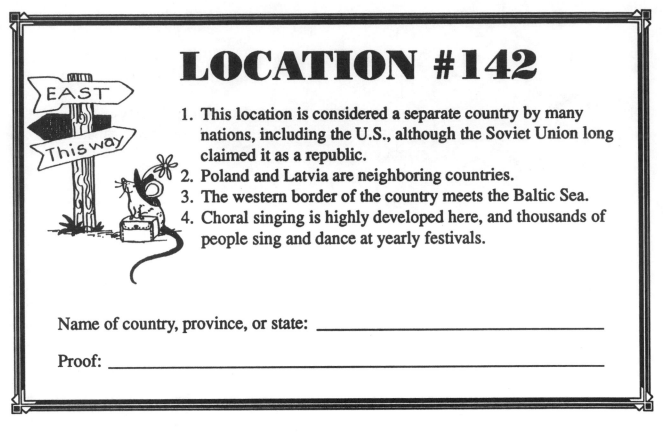

LOCATION #142

1. This location is considered a separate country by many nations, including the U.S., although the Soviet Union long claimed it as a republic.
2. Poland and Latvia are neighboring countries.
3. The western border of the country meets the Baltic Sea.
4. Choral singing is highly developed here, and thousands of people sing and dance at yearly festivals.

Name of country, province, or state: _____

Proof: _____

Creativity Across The Curriculum

1. Ancient songs called dainos, as well as folk tales, have been handed down from generation to generation. These songs and stories tell of the culture and history of the people. Create a song or story that tells of your family's cultural background or history.

2. People here have been discouraged from pursuing religious practices. Few priests have been trained to replace those who retire, and religious correspondence is frowned upon. Create a code that would allow you to write to someone without government officials understanding what is being said. Write a message to a friend using your new code.

LOCATION #143

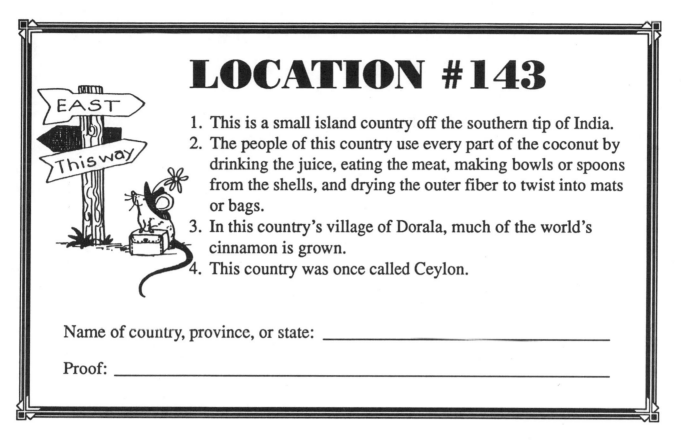

1. This is a small island country off the southern tip of India.
2. The people of this country use every part of the coconut by drinking the juice, eating the meat, making bowls or spoons from the shells, and drying the outer fiber to twist into mats or bags.
3. In this country's village of Dorala, much of the world's cinnamon is grown.
4. This country was once called Ceylon.

Name of country, province, or state: _____

Proof: _____

Creativity Across The Curriculum

1. Finish each of the following sentences:
   • If the world did not have rubber trees . . .
   • Drinking tea is like . . .
   • A food that is good with cinnamon is . . .
   • Living on an island would be like . . .
   • If we had to produce all of our own food, I would . . .
   • A mammoty is . . .
   • If I had to walk two miles to school . . .
   • If we had only coconut milk to drink . . .
   • Jack fruit is . . .
   • Coconuts make great . . .

2. Create a ten-line nonsense poem about life on an island.

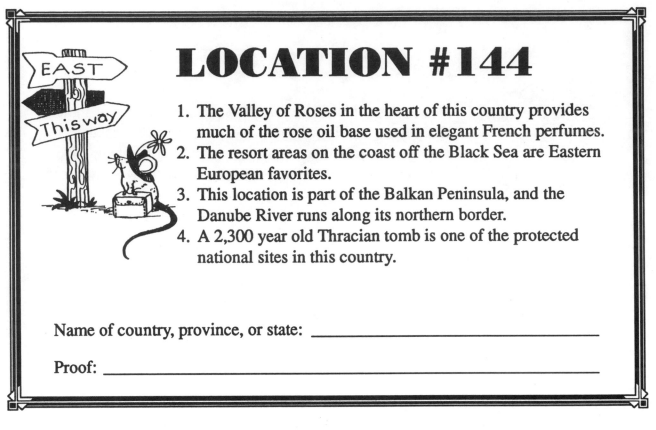

LOCATION #144

1. The Valley of Roses in the heart of this country provides much of the rose oil base used in elegant French perfumes.
2. The resort areas on the coast off the Black Sea are Eastern European favorites.
3. This location is part of the Balkan Peninsula, and the Danube River runs along its northern border.
4. A 2,300 year old Thracian tomb is one of the protected national sites in this country.

Name of country, province, or state: _____

Proof: _____

Creativity Across The Curriculum

1. The name of a perfume or cologne helps attract customers. Brainstorm a list of five new names for perfumes and five new names for colognes that you think would attract the public's attention.

2. Create a pair of designer sunglasses that you could wear to one of Location Number 144's beach resorts. Draw them on tag board and cut them out. Model them for a friend.

3. Do research to find out how the Black Sea got its name. If you could rename it, what would you call it?

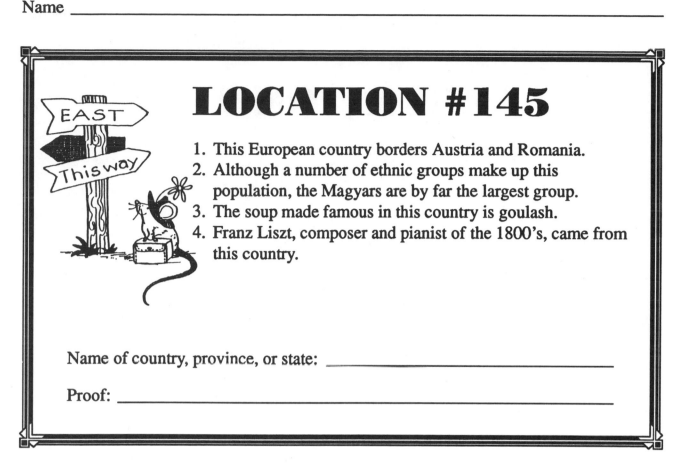

LOCATION #145

1. This European country borders Austria and Romania.
2. Although a number of ethnic groups make up this population, the Magyars are by far the largest group.
3. The soup made famous in this country is goulash.
4. Franz Liszt, composer and pianist of the 1800's, came from this country.

Name of country, province, or state: _____

Proof: _____

Creativity Across The Curriculum

1. Create a web of facts about this country using only pictures. The pictures may be drawn or cut from magazines.

2. Pretend you are a student living in Location Number 145. Write a fictitious letter to your aunt and uncle who live in California convincing them that they should move to this country from the U.S.

3. Create a book jacket to advertise a nonfiction book about this country.

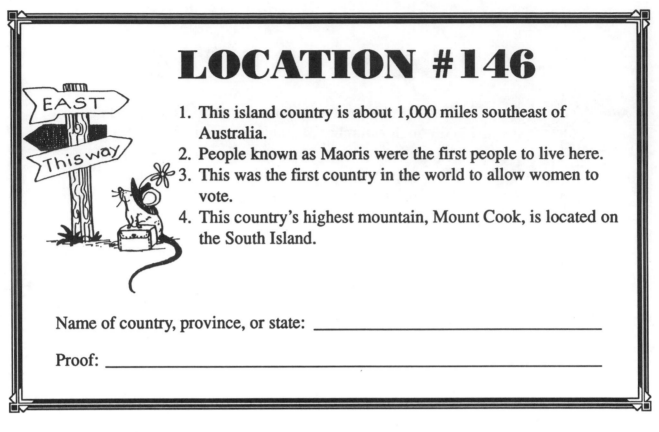

LOCATION #146

1. This island country is about 1,000 miles southeast of Australia.
2. People known as Maoris were the first people to live here.
3. This was the first country in the world to allow women to vote.
4. This country's highest mountain, Mount Cook, is located on the South Island.

Name of country, province, or state: _____

Proof: _____

Creativity Across The Curriculum

1. This country has no snakes. List five reasons to be sad about this fact.

2. List six sources where you could find additional information about this country.

3. Compare three of this country's native birds to the extinct moa.

4. Brainstorm a list of ten questions that you would like to ask about this country. Answer one of them.

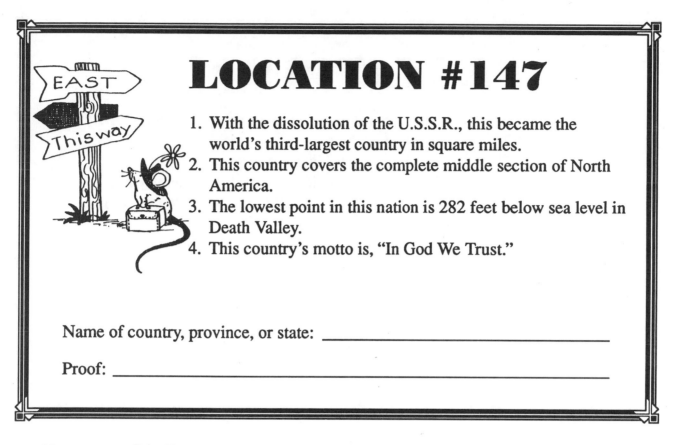

LOCATION #147

1. With the dissolution of the U.S.S.R., this became the world's third-largest country in square miles.
2. This country covers the complete middle section of North America.
3. The lowest point in this nation is 282 feet below sea level in Death Valley.
4. This country's motto is, "In God We Trust."

Name of country, province, or state: _____

Proof: _____

Creativity Across The Curriculum

1. "The Star-Spangled Banner" was adopted as the national anthem in 1931. It is often criticized as being too difficult for most people to sing. If you could choose a new national anthem, what would you choose? Why?

2. Urban and rural life are very different from each other in this country. Which do you prefer? List ten reasons for your choice.

3. There are close to 9,000 libraries in this nation. If you were in charge of opening a new library in your town, what are the first twenty-five books that you would want to order?

Name _____

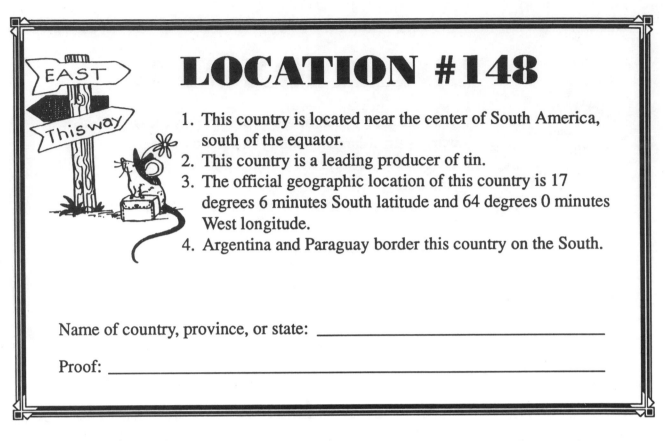

LOCATION #148

1. This country is located near the center of South America, south of the equator.
2. This country is a leading producer of tin.
3. The official geographic location of this country is 17 degrees 6 minutes South latitude and 64 degrees 0 minutes West longitude.
4. Argentina and Paraguay border this country on the South.

Name of country, province, or state: _____

Proof: _____

Creativity Across The Curriculum

1. Tin mining is an important industry here. Write an editorial warning the public about the safety hazards of mining.

2. In parts of this country, tropical rain forests flourish. More species of trees are found in the rain forest than anywhere else on earth. In a 2.5-acre area of rain forest, scientists have identified 179 species. A comparable section of forest in the U.S. has about seven species. Compare the appearance of the following trees—brazil-woods, mahoganies, kapoks, and strangler trees. Put together a chart that includes the following information:

Type of Tree	Height	Notable Features	Uses for Humankind

Name _____

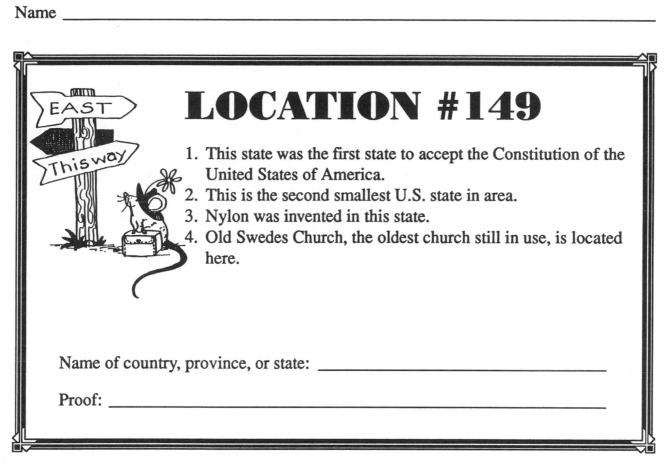

LOCATION #149

1. This state was the first state to accept the Constitution of the United States of America.
2. This is the second smallest U.S. state in area.
3. Nylon was invented in this state.
4. Old Swedes Church, the oldest church still in use, is located here.

Name of country, province, or state: _____

Proof: _____

Creativity Across The Curriculum

1. People are curious about many things. Interview your classmates to find out what they are most curious about and how they would find an answer to a question they might have. Then create a graph to show the most common sources of information consulted when answering questions. These sources might include: reference books, authority figures, friends, experiments, and parents or family.

2. Many colonial homes in this state are still occupied today. List some differences between colonial homes and our modern-day homes.

3. Select the region of the United States in which you would most like to live. Create a pop-up picture of a scene from this area.

Name _____

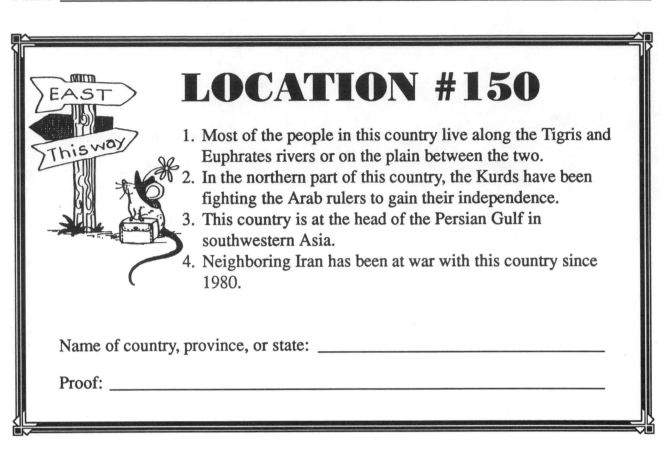

LOCATION #150

1. Most of the people in this country live along the Tigris and Euphrates rivers or on the plain between the two.
2. In the northern part of this country, the Kurds have been fighting the Arab rulers to gain their independence.
3. This country is at the head of the Persian Gulf in southwestern Asia.
4. Neighboring Iran has been at war with this country since 1980.

Name of country, province, or state: _____

Proof: _____

Creativity Across The Curriculum

1. This area was once part of a land called Mesopotamia. Order these tributes to the past according to their date of construction.

_____ Taj Mahal, India
_____ St. Basil's Cathedral, Moscow
_____ The Suez Canal, Egypt
_____ The Great Wall of China
_____ The Great Pyramid at Giza, Egypt
_____ Lascaux Cave Paintings, France
_____ Eiffel Tower, France
_____ The Sphinx, Egypt
_____ Statue of Liberty, United States of America

2. Which eight nations border the Persian Gulf? Prepare a graph that shows the three leading oil producers and the amount of oil each produced in a recent year.

Name _____

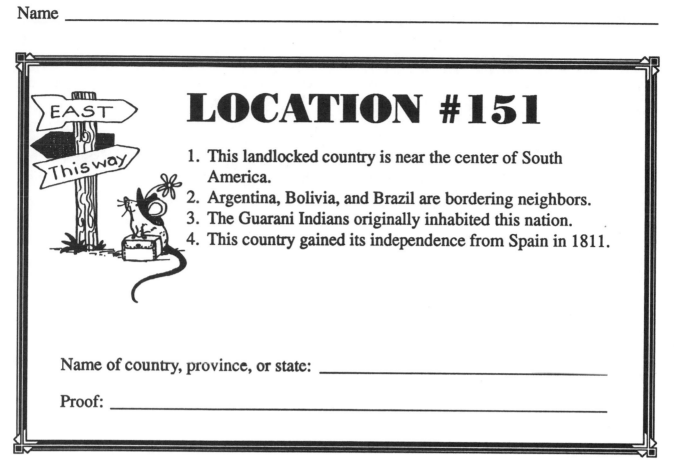

LOCATION #151

1. This landlocked country is near the center of South America.
2. Argentina, Bolivia, and Brazil are bordering neighbors.
3. The Guarani Indians originally inhabited this nation.
4. This country gained its independence from Spain in 1811.

Name of country, province, or state: _____

Proof: _____

Creativity Across The Curriculum

1. After researching the culture of the Guarani Indians, write a poem to describe their way of life.

2. The most famous handicraft of this nation is nandutì lace. It is made by the women of Itauguà. The spider web lace is made into complicated patterns of flowers, animals, and familiar objects. Sketch a lace pattern that represents something valued by your culture.

3. Sopa paraguaya is often served on special occasions. List ten foods that are served by your family on special occasions.

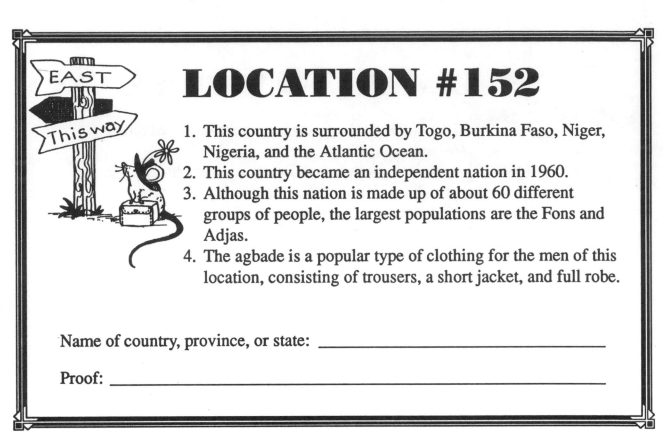

LOCATION #152

1. This country is surrounded by Togo, Burkina Faso, Niger, Nigeria, and the Atlantic Ocean.
2. This country became an independent nation in 1960.
3. Although this nation is made up of about 60 different groups of people, the largest populations are the Fons and Adjas.
4. The agbade is a popular type of clothing for the men of this location, consisting of trousers, a short jacket, and full robe.

Name of country, province, or state: _____

Proof: _____

Creativity Across The Curriculum

1. Brainstorm a list of adjectives that would describe life in this country.

2. Predict ten changes that you could imagine taking place in this country during the next 25 years.

3. Create a detailed, drawn-to-scale map of the continent on which this country is located.

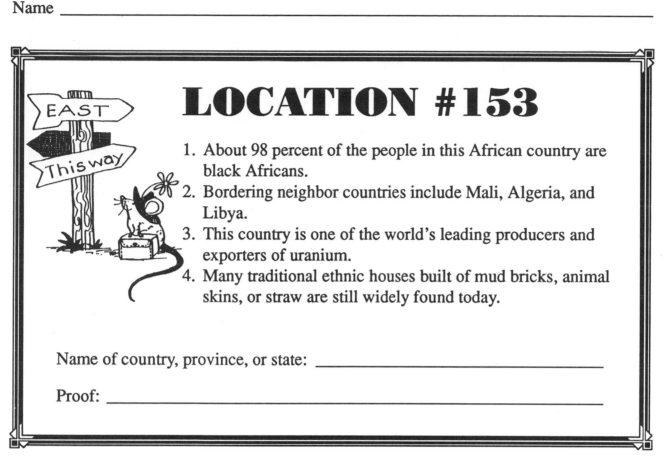

LOCATION #153

1. About 98 percent of the people in this African country are black Africans.
2. Bordering neighbor countries include Mali, Algeria, and Libya.
3. This country is one of the world's leading producers and exporters of uranium.
4. Many traditional ethnic houses built of mud bricks, animal skins, or straw are still widely found today.

Name of country, province, or state: _____

Proof: _____

Creativity Across The Curriculum

1. Translator phones are being perfected that will allow people who speak different languages to communicate with each other. Their conversations will be instantly translated from the caller's language to the listener's language. Choose three countries where you would most want to talk to students your own age. What countries would you call, and what questions would you ask?

2. The roads in this country are few and poorly paved. As you travel along a narrow strip of roadway, you suddenly come upon an elephant in your path. Due to the large ruts and boulders off the roadway, you cannot go around it. What will you do?

Name _____

LOCATION #154

EAST
This way

1. This country is composed of about 150 South Pacific Ocean islands.
2. When Captain James Cook explored these islands in 1773, he called them the "Friendly Islands."
3. Honolulu, Hawaii, is about 3,000 miles northeast of these islands.
4. The largest island is Tongatapu.

Name of country, province, or state: _____

Proof: _____

Creativity Across The Curriculum

1. People here fish for and eat shark. What type of shark is most commonly found in the waters surrounding these islands? Prepare a large diagram of one type of shark found here. Inside the diagram write eight to ten facts about the shark. Use complete sentences.

2. Cyclones may occur here. Prepare an emergency procedures booklet for people who may be facing such a storm.

3. The main islands are largely coral reefs. Draw and diagram a coral reef. How does it support life?

LOCATION #155

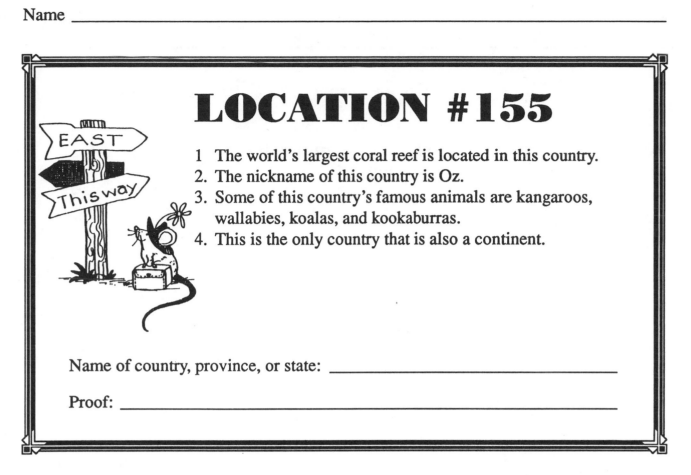

1 The world's largest coral reef is located in this country.
2. The nickname of this country is Oz.
3. Some of this country's famous animals are kangaroos, wallabies, koalas, and kookaburras.
4. This is the only country that is also a continent.

Name of country, province, or state: _____

Proof: _____

Creativity Across The Curriculum

1. Create teams with captains. Assemble eight small boxes numbered from one through eight. Into each box put questions concerning a country that you have chosen. The questions should range from very easy (one point) to difficult (eight points). Each team is responsible for contributing three questions to each box. The answers should be written on the back of the questions. When the boxes are complete, team captains begin selecting boxes in turn and then work with team members to answer each question. When a question is answered correctly, the team answering it is awarded the number of points on the box. The winning team is the team that accumulates the most points.

2. Find a way to determine the perimeter of this country. Compare it to the perimeter of your state.

Discover and Create

Reptiles such as the frilled lizard and the thorny devil roam this country. You are part of a team of scientists studying reptilian life on the continent. Describe a new reptile discovered by your team. Study the reptile closely in order to present as many facts as possible in your scientific report. What does it eat? Where does it prefer to live? What are its daily habits? What does it look like? Who are its enemies? Then draw and color a large picture of your discovery.

Birdy Birdy in the Sky?

Location Number 155 is a land of many unusual birds, such as the emu. Create your own unusual bird, using a variety of types of paper rolled into cones and cylinders. Sculpt the bird in the third dimension by following these simple directions:

1. The base of the bird will be created by rolling an eight-inch square piece of heavy paper into a cone. Roll two three-inch pieces of heavy tagboard into tight cylinders to create the feet. Punch two holes in the base cone and attach the feet.

2. Using a six-inch piece of paper, create a cone for the head of the bird. Attach this to the body with another tightly wound cylinder of heavy tagboard that is five inches square.

3. The basic body of the bird is now complete. Add brightly colored feathers made from different colors of paper, paint where needed, and add any additional materials that you think would add interest to your creation.

Be original. Be inventive. The sky is the limit!

Crayon Rubbings

Country 155 has thousands of beautiful wildflowers. Create your own impression of a desert that has suddenly come to life with the heavy rains. To create your wildflowers, follow the directions for crayon rubbings. After you have completed your rubbings, cut them into floral shapes and glue them onto a large buff- or tan-colored piece of paper. You may also choose to add desert animals to your composition. Add necessary detail with crayon or cut paper.

Directions for crayon rubbings:

On your desk, arrange an assortment of cut pieces of heavy paper, scraps of fabric such as burlap, netting, and string of several sizes. If you have other textured scraps, such as textured wallpaper, add these to your assortment.

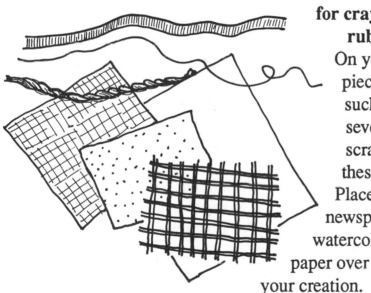

Place a piece of newsprint or watercolor paper over your creation.

Remove the wrapping from your crayon and hold it firmly on its side, flat against the paper. Press firmly and evenly as you color the paper. You will wish to use several colors. You may even choose to overlap colors. Once the coloring is complete, cut your floral shapes and glue your picture together.

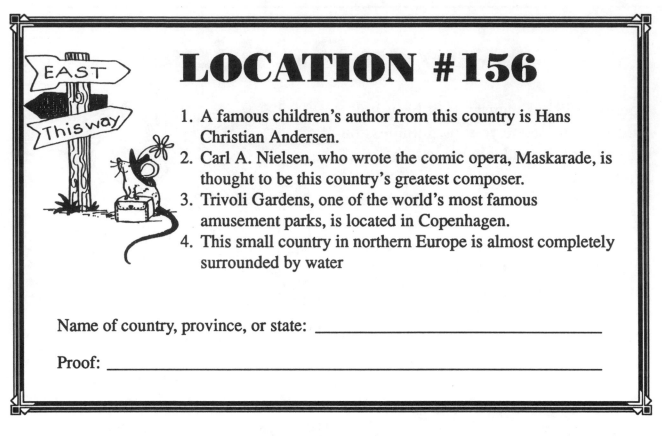

LOCATION #156

1. A famous children's author from this country is Hans Christian Andersen.
2. Carl A. Nielsen, who wrote the comic opera, Maskarade, is thought to be this country's greatest composer.
3. Trivoli Gardens, one of the world's most famous amusement parks, is located in Copenhagen.
4. This small country in northern Europe is almost completely surrounded by water

Name of country, province, or state: _____

Proof: _____

Creativity Across The Curriculum

1. Retell one of Andersen's tales by completely modernizing it. For example, "The Ugly Duckling" might have a rock star as the main character experiencing a "bad hair day."

2. Folk high schools give high school students a general education in government, history, and literature. Each course may last up to six months and the students live at the schools. If you could design a school from scratch, what would you want to include? How would it be different from your school?

• Name Of My Private School: _____

• Classes offered: _____

• Special features: _____

• Ages of students: _____

• Who could attend? _____

• How would it better prepare students for their future? _____

_____

_____

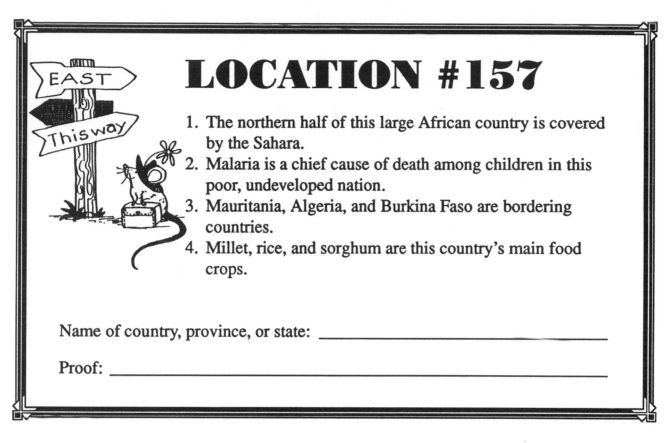

LOCATION #157

1. The northern half of this large African country is covered by the Sahara.
2. Malaria is a chief cause of death among children in this poor, undeveloped nation.
3. Mauritania, Algeria, and Burkina Faso are bordering countries.
4. Millet, rice, and sorghum are this country's main food crops.

Name of country, province, or state: _____

Proof: _____

Creativity Across The Curriculum

1. Create a bookjacket that will advertise a nonfiction book about the continent of Africa.

2. From magazines, cut out the heads, legs, and bodies of various types of animals, then divide them among three boxes. Draw out pieces until you have one body, one head and four legs. Create a story about your new animal.

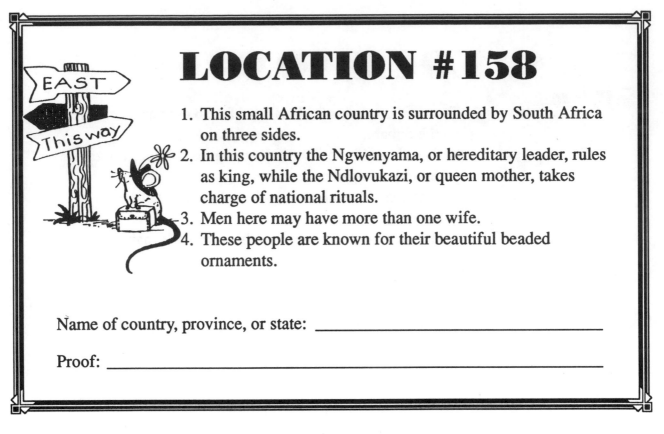

LOCATION #158

1. This small African country is surrounded by South Africa on three sides.
2. In this country the Ngwenyama, or hereditary leader, rules as king, while the Ndlovukazi, or queen mother, takes charge of national rituals.
3. Men here may have more than one wife.
4. These people are known for their beautiful beaded ornaments.

Name of country, province, or state: _____

Proof: _____

Creativity Across The Curriculum

1. All of the men here are organized into different age groups. Each age group has a different role in special ceremonies. Create age groups for the men and women in the U.S. Describe the job that each group would have in putting together a July Fourth Celebration for their town.

2. Kaolin (clay used for pottery) is found in this country. Create a clay pot to honor the craftspeople of this country.

3. This nation is one of the few African countries that exports more than it imports. How does this affect a country's economy?

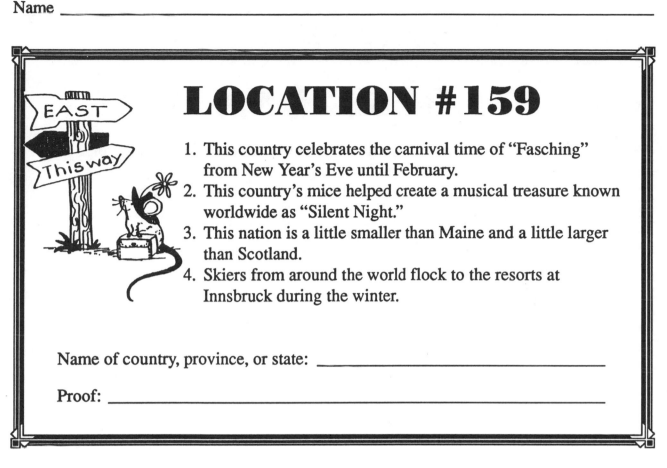

LOCATION #159

1. This country celebrates the carnival time of "Fasching" from New Year's Eve until February.
2. This country's mice helped create a musical treasure known worldwide as "Silent Night."
3. This nation is a little smaller than Maine and a little larger than Scotland.
4. Skiers from around the world flock to the resorts at Innsbruck during the winter.

Name of country, province, or state: _____

Proof: _____

Creativity Across The Curriculum

1. If bicycles were a major means of transportation in the U.S., as they are in this country, how would they change? Draw a picture of the bicycles you would design for the members of your family.

2. Tourism allows this location to make a fair amount of money. Design a brochure that could be used to promote the tourist industry of Location Number 159.

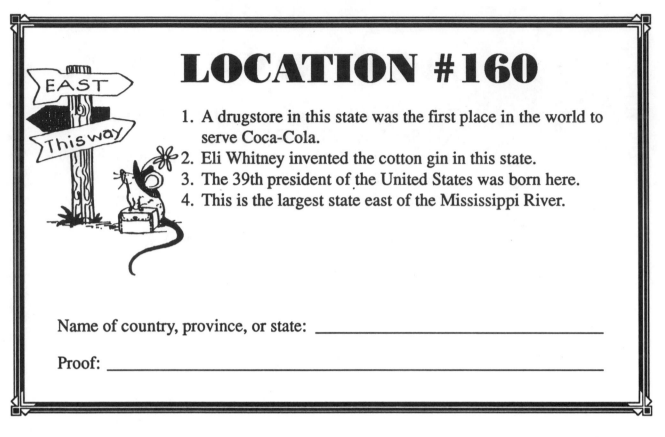

LOCATION #160

1. A drugstore in this state was the first place in the world to serve Coca-Cola.
2. Eli Whitney invented the cotton gin in this state.
3. The 39th president of the United States was born here.
4. This is the largest state east of the Mississippi River.

Name of country, province, or state: _____

Proof: _____

Creativity Across The Curriculum

1. The invention of the cotton gin has been important to the development of American industry. List products that have resulted from this invention.

2. The American family is changing. How has the organization *The Girl Scouts of the United States of America* also changed in the past fifty years? This organization was founded in 1912 in Location Number 160 by Juliette Lowe.

3. This state produces more peanuts than any other state. Do research to create a list of twenty products made from peanuts.

LOCATION #161

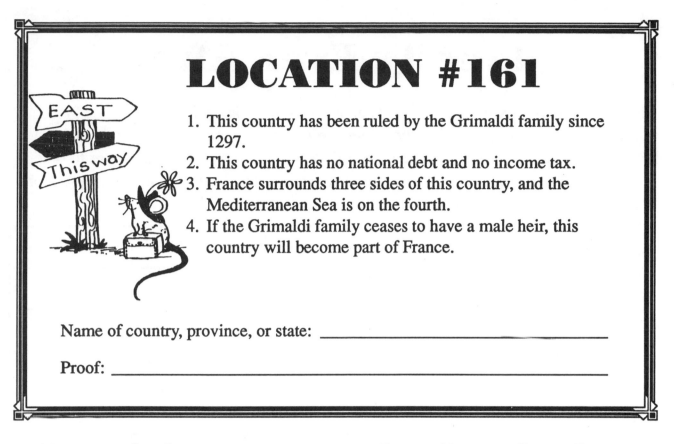

1. This country has been ruled by the Grimaldi family since 1297.
2. This country has no national debt and no income tax.
3. France surrounds three sides of this country, and the Mediterranean Sea is on the fourth.
4. If the Grimaldi family ceases to have a male heir, this country will become part of France.

Name of country, province, or state: _____

Proof: _____

Creativity Across The Curriculum

1. Prince Rainier III married American movie actress Grace Kelly here in 1956. She then became Princess Grace. You have been given the opportunity to become an actor/actress or Prince/Princess. Which "job" would you prefer? Write a letter to your diary explaining your choice.

2. The annual Grand Prix Auto Race in this country features the greatest racecar drivers from around the world. Design the car that you would drive if you were part of this race.

Happily Ever After

Living in Location Number 161 has been compared to living in a fairy tale. Choose five fairy-tale characters. Using the survey below, determine if they were leading happy lives. Rate each one on a scale from one to ten. Then total each score to determine if each fairy-tale life was a happy one. Write a description of the happiest character that you find in your survey.

Name of Character	Healthy	Safe	Kind	Happy	Loved	Free	**Totals**

_____

_____

_____

_____

_____

_____

_____

_____

Name _____

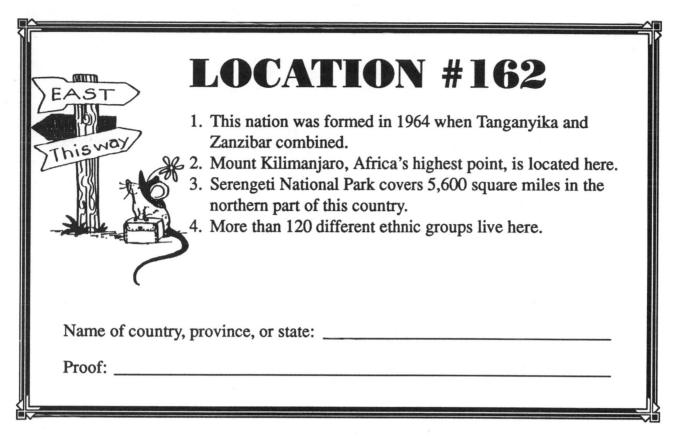

LOCATION #162

1. This nation was formed in 1964 when Tanganyika and Zanzibar combined.
2. Mount Kilimanjaro, Africa's highest point, is located here.
3. Serengeti National Park covers 5,600 square miles in the northern part of this country.
4. More than 120 different ethnic groups live here.

Name of country, province, or state: _____

Proof: _____

Creativity Across The Curriculum

1. Along the African coast one may see a look-alike of an old Arab sailing vessel known as a dhow. This ship has served as a model for many ships over the centuries. Draw a model of a sailing vessel that could be used today to transport goods from one country to another.

2. Much of the world's supply of cloves comes from this nation's island of Zanzibar. Find a recipe that uses cloves. Taste cloves. Describe the flavor.

LOCATION #163

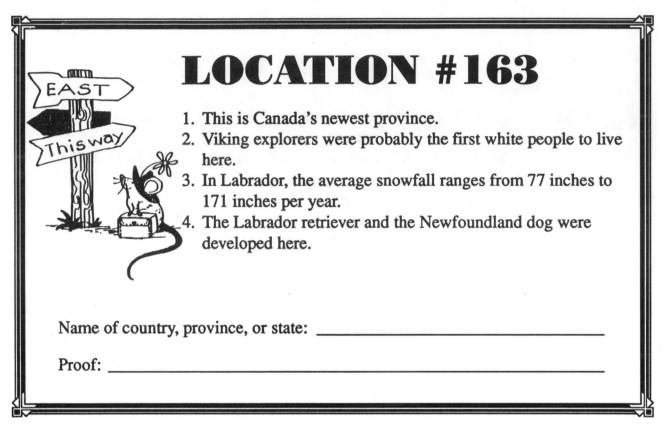

1. This is Canada's newest province.
2. Viking explorers were probably the first white people to live here.
3. In Labrador, the average snowfall ranges from 77 inches to 171 inches per year.
4. The Labrador retriever and the Newfoundland dog were developed here.

Name of country, province, or state: _____

Proof: _____

Creativity Across The Curriculum

1. If someone left a tiny Labrador retriever puppy on your doorstep, how would you convince your family to keep it? Develop five convincing arguments that you could use.

2. The answer is Vikings. List as many questions as you can that result in this answer.

3. How is a Labrador retriever like a Newfoundland dog? Sketch each one.

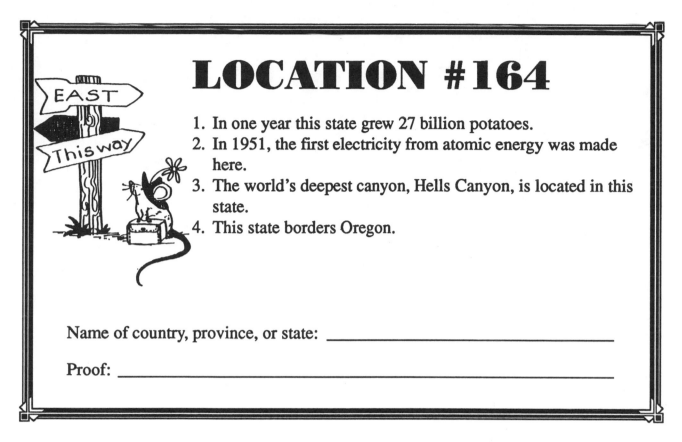

LOCATION #164

1. In one year this state grew 27 billion potatoes.
2. In 1951, the first electricity from atomic energy was made here.
3. The world's deepest canyon, Hells Canyon, is located in this state.
4. This state borders Oregon.

Name of country, province, or state: _____

Proof: _____

Creativity Across The Curriculum

1. List as many ways as you can think of to prepare potatoes.

2. Describe a mathematical process that could be used to measure the depth of Hells Canyon from the upper rim.

3. Do some research to find out how the Snake River got its name.

4. Imagine that potatoes are available in a variety of colors. What flavor would each color be? How would this affect the potato industry?

LOCATION #165

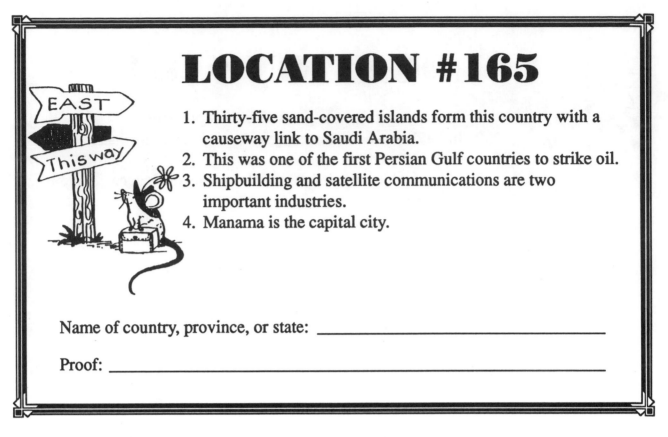

1. Thirty-five sand-covered islands form this country with a causeway link to Saudi Arabia.
2. This was one of the first Persian Gulf countries to strike oil.
3. Shipbuilding and satellite communications are two important industries.
4. Manama is the capital city.

Name of country, province, or state: _____

Proof: _____

Creativity Across The Curriculum

1. Country Number 165 is noted for processing aluminum. List ten uses for aluminum.

2. Animals large and small have ways to defend and protect themselves. How do the animals of the sea protect themselves? Choose three to research.

3. Write an answer for the following "What if" question.
What if all the oil in the world suddenly disappeared?

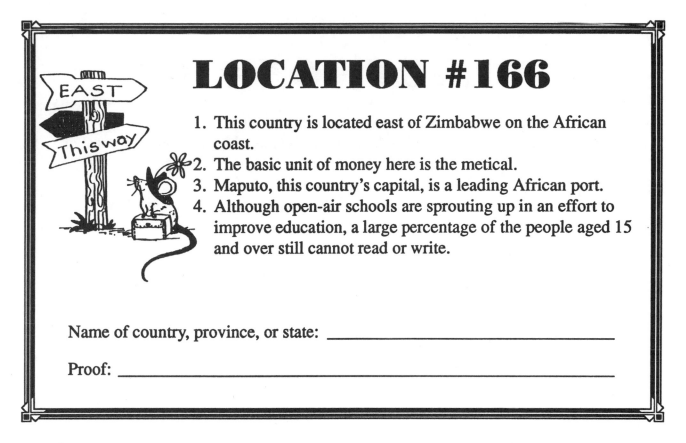

LOCATION #166

1. This country is located east of Zimbabwe on the African coast.
2. The basic unit of money here is the metical.
3. Maputo, this country's capital, is a leading African port.
4. Although open-air schools are sprouting up in an effort to improve education, a large percentage of the people aged 15 and over still cannot read or write.

Name of country, province, or state: _____

Proof: _____

Creativity Across The Curriculum

1. Most people who live in Location Number 166 are black Africans. Do some research about a famous black *American*. Describe the hardships he or she has faced during his or her lifetime. How does this person's life compare to yours?

2. Write a message of hope and kindness to share with a friend. Place it in a paper fortune cookie and pass it on.

3. Create a rebus story using animals found in Africa. A rebus is a story that is told partly with words and partly with pictures.

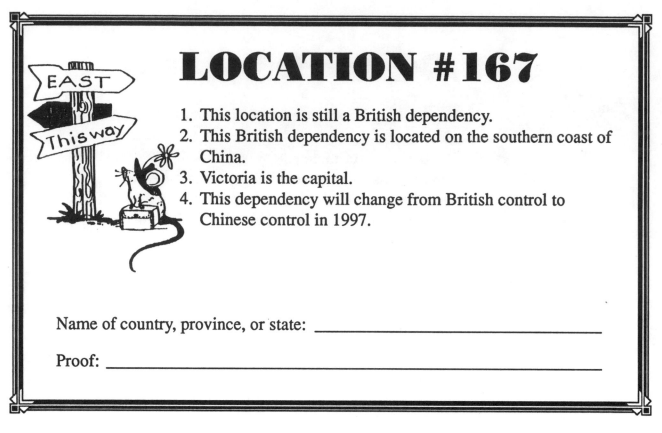

LOCATION #167

1. This location is still a British dependency.
2. This British dependency is located on the southern coast of China.
3. Victoria is the capital.
4. This dependency will change from British control to Chinese control in 1997.

Name of country, province, or state: _____

Proof: _____

Creativity Across The Curriculum

1. Much clothing is imported by the United States from this location. Do research to find out some of the brand names of clothing that is manufactured in Location Number 167.

2. Although this location is nearly surrounded by water, its dry winters force its people to buy millions of gallons of water from China each year. What are some ways that people here could avoid having to buy so much water?

3. Add to the following web:

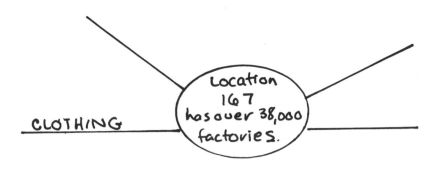

CLOTHING

Location 167 has over 38,000 factories.

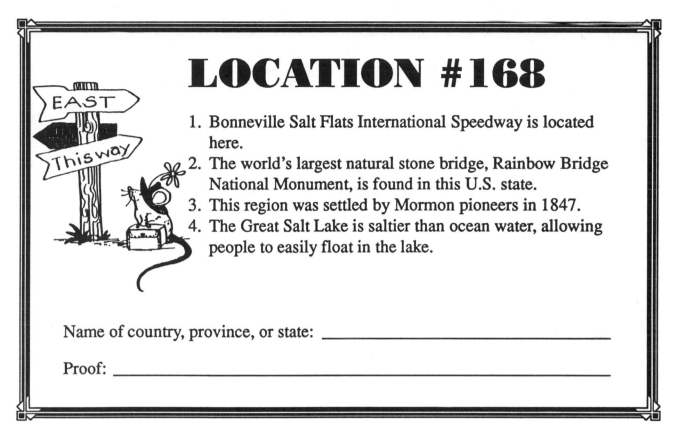

LOCATION #168

1. Bonneville Salt Flats International Speedway is located here.
2. The world's largest natural stone bridge, Rainbow Bridge National Monument, is found in this U.S. state.
3. This region was settled by Mormon pioneers in 1847.
4. The Great Salt Lake is saltier than ocean water, allowing people to easily float in the lake.

Name of country, province, or state: _____

Proof: _____

Creativity Across The Curriculum

1. The Great Salt Lake is one of the seven natural wonders of the world. Do some research to find out what the other natural wonders are. List them. Which one would you most like to visit? Why?

2. Compare the state bird of this state with the state bird of a bordering state. Which bird is the most beautiful? Support your answer.

3. Write to the Division of Travel Development in this state to request information on tourist attractions.

LOCATION #169

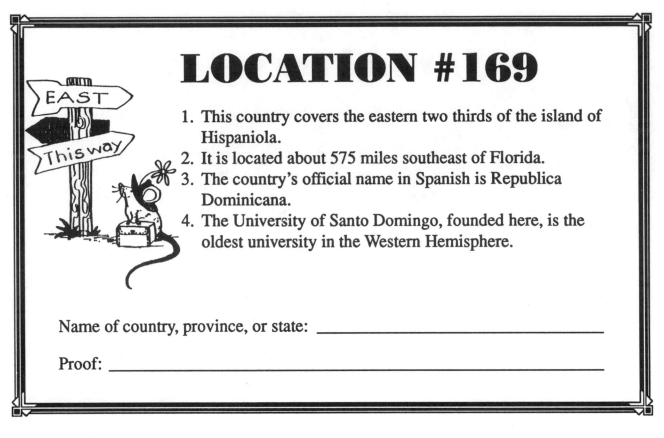

1. This country covers the eastern two thirds of the island of Hispaniola.
2. It is located about 575 miles southeast of Florida.
3. The country's official name in Spanish is Republica Dominicana.
4. The University of Santo Domingo, founded here, is the oldest university in the Western Hemisphere.

Name of country, province, or state: _____

Proof: _____

Creativity Across The Curriculum

1. Avocados and sugar cane are two important crops here. For one week you have nothing but avocados, sugar cane, and water. What will you do? List your typical meals for one day.

2. On a visit to this country, you break out in a rash from head to toe. There are no doctors available to see you. Do some research and list the possible reasons for your rash and the appropriate treatments. What do you feel is the most likely reason for this red, itchy rash?

LOCATION #170

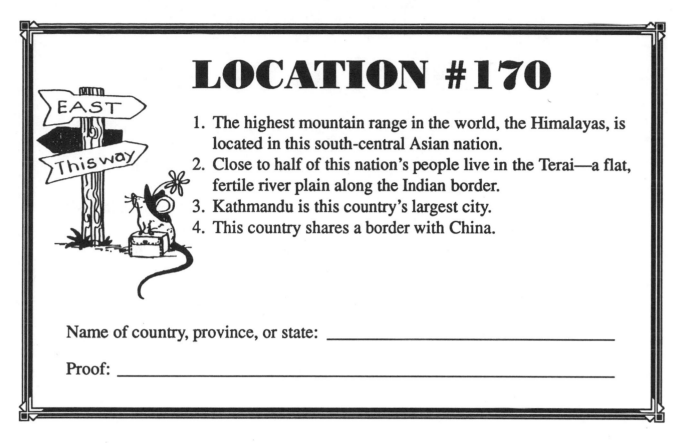

1. The highest mountain range in the world, the Himalayas, is located in this south-central Asian nation.
2. Close to half of this nation's people live in the Terai—a flat, fertile river plain along the Indian border.
3. Kathmandu is this country's largest city.
4. This country shares a border with China.

Name of country, province, or state: _____

Proof: _____

Creativity Across The Curriculum

1. What is the difference in height between Mount Everest and Mount McKinley?
   - If it took ten minutes to climb 290 feet, how long would it take to climb each mountain?
   - How much taller is Everest than Kilimanjaro?

2. As you open your door one morning, you see a news crew outside with a large banner saying that you have just inherited the palace of Country Number 170 and have been chosen as its king or queen. What are the first ten things that you will do? In order to maintain your title, you must live in Location Number 170. Are you willing to do so? You may bring ten people with you. Who will you bring if you go?

Name _____

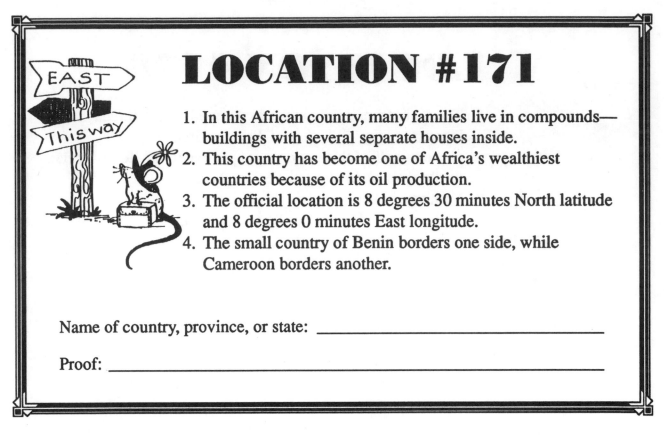

LOCATION #171

1. In this African country, many families live in compounds—buildings with several separate houses inside.
2. This country has become one of Africa's wealthiest countries because of its oil production.
3. The official location is 8 degrees 30 minutes North latitude and 8 degrees 0 minutes East longitude.
4. The small country of Benin borders one side, while Cameroon borders another.

Name of country, province, or state: _____

Proof: _____

Creativity Across The Curriculum

1. You are in charge of airline flight planning. Part of your job is to determine the distances from one place to another. Calculate direct-path mileage to the city of Lagos from each of the following ports of departure:
   - Venice
   - New York City
   - Hong Kong
   - Mexico City
   - Lima
   - Melbourne

2. Both an almanac and an atlas can provide helpful information. How are these references alike? Different? Compare yourself to either an atlas or an almanac.

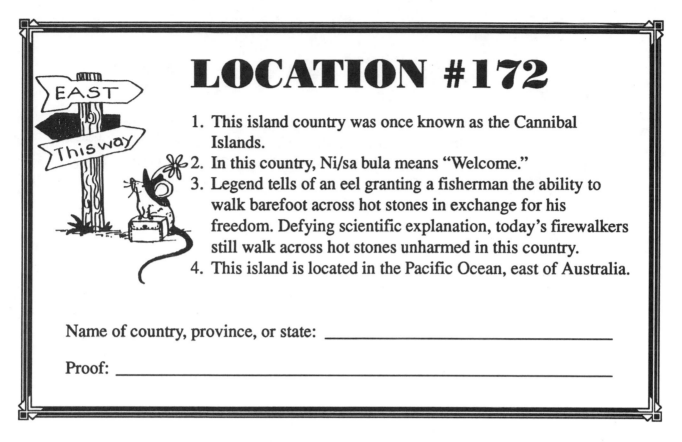

LOCATION #172

1. This island country was once known as the Cannibal Islands.
2. In this country, Ni/sa bula means "Welcome."
3. Legend tells of an eel granting a fisherman the ability to walk barefoot across hot stones in exchange for his freedom. Defying scientific explanation, today's firewalkers still walk across hot stones unharmed in this country.
4. This island is located in the Pacific Ocean, east of Australia.

Name of country, province, or state: _____

Proof: _____

Creativity Across The Curriculum

• Create a list of twenty-five likely animals of the land and sea that would be associated with an island country.

• Design and build a scale model island. Provide a key.

• Much of our earth's water is polluted. Predict what our water will be like in twenty years if pollution continues at its current rate. What changes will this cause on earth?

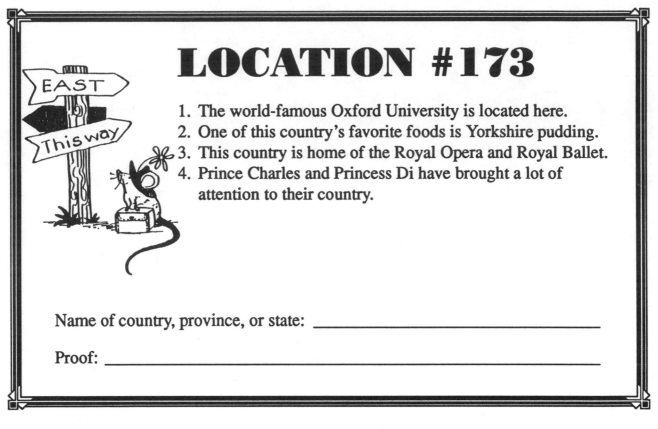

LOCATION #173

1. The world-famous Oxford University is located here.
2. One of this country's favorite foods is Yorkshire pudding.
3. This country is home of the Royal Opera and Royal Ballet.
4. Prince Charles and Princess Di have brought a lot of attention to their country.

Name of country, province, or state: _____

Proof: _____

Creativity Across The Curriculum

1. Although the people here speak English, many of their words are quite different from ours. For example, a truck is a lorry and an apartment is a flat. List ten of our English words that would seem strange to people in other English-speaking countries.

2. Read several folktales (you may use *British Folk Tales* by Kevin Crossley Holland, Orchard Books, New York, 1988). Using a British folktale, prepare yourself to be the storyteller for a group of younger children. Create a prop to use during your presentation.

The Marks Of A Culture

Big Ben is a noted historic timepiece and landmark. Design a clock for your city that could become an historic monument. Explain the significance of your design.

Cricket is the number one sporting event in this country. Create a rules and instruction book for Cricket that could be used by a coach in the U.S. who wished to teach a young group of children how to play the game.

Gather a group of two or three friends together and read aloud a portion of one of Shakespeare's great works such as "Romeo and Juliet." After reading for fifteen minutes, discuss how you feel about the work of Shakespeare. How is it different from other things that you have read? Why do you think this work has been popular for so many years?

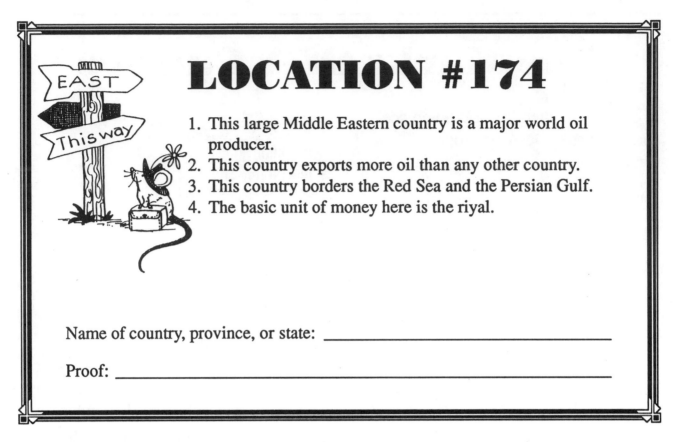

LOCATION #174

1. This large Middle Eastern country is a major world oil producer.
2. This country exports more oil than any other country.
3. This country borders the Red Sea and the Persian Gulf.
4. The basic unit of money here is the riyal.

Name of country, province, or state: _____

Proof: _____

Creativity Across The Curriculum

1. The Bedouins are a culture of nomadic herders who live in temporary desert campsites and roam from place to place with their camels, goats, and sheep. List twenty-five ways in which your lifestyle differs from theirs.

2. Boys and girls here attend separate schools. List the pluses and minuses, and the interesting facts and feelings, about all-boy and all-girl schools. Use this information to help you decide whether or not a similar arrangement would work in the United States.

Example:

+		Interesting
Girls wouldn't worry about competing with boys in math and science.	There wouldn't be boys to dance with during gym dance lessons.	Would the teachers all be men or all be women?

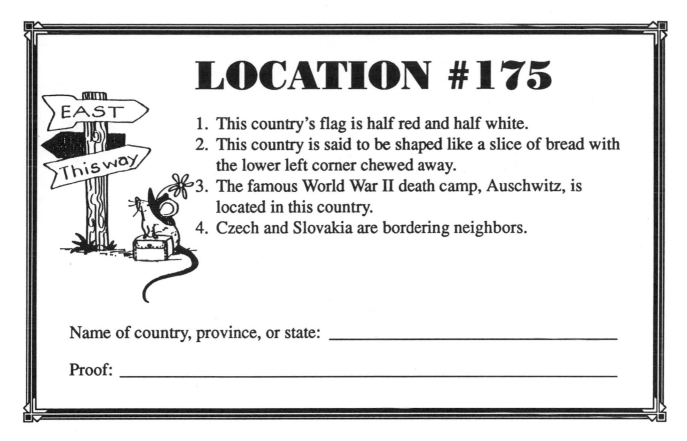

LOCATION #175

1. This country's flag is half red and half white.
2. This country is said to be shaped like a slice of bread with the lower left corner chewed away.
3. The famous World War II death camp, Auschwitz, is located in this country.
4. Czech and Slovakia are bordering neighbors.

Name of country, province, or state: _____

Proof: _____

Creativity Across The Curriculum

1. If you wished to run for president of a country, what would you use as your "political platform"? List your top three ideas and explain each one.

2. Potatoes are the chief food product of this country. You are the head writer for an advertising company hired to promote potatoes. Create a mini-billboard on 12" x 18" white paper that shows your potato advertisement.

3. A Martian has just moved in next door. You don't want him to scare the neighbors, so you teach him as much as possible about planet Earth. He can read and remember ten pages of information per day. Which ten pages of the almanac would you have him read first?

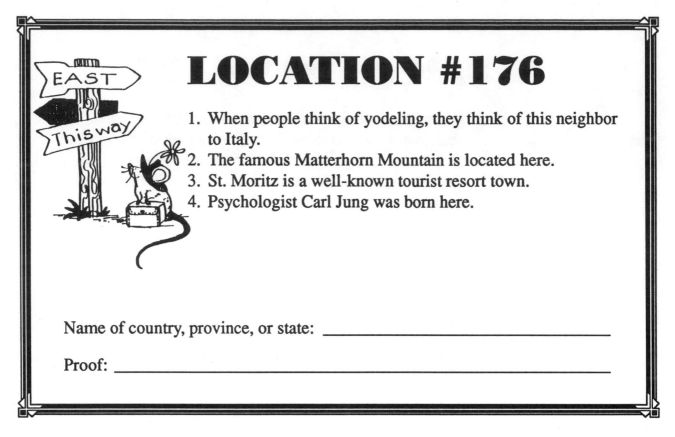

LOCATION #176

1. When people think of yodeling, they think of this neighbor to Italy.
2. The famous Matterhorn Mountain is located here.
3. St. Moritz is a well-known tourist resort town.
4. Psychologist Carl Jung was born here.

Name of country, province, or state: _____

Proof: _____

Creativity Across The Curriculum

1. The Santa Claus from this country, known as Samichlaus, doesn't use reindeer. If the Santa Claus of the North Pole no longer used reindeer, what would he use? Write a story about Santa Claus making his deliveries without reindeer.

2. About 100 million timepieces or clock and watch mechanisms are made here each year:
   a) If 200 different countries, states, and provinces were each sent an equal number of timepieces or parts each year, how many would each receive?
   b) If the craftsworkers creating these timepieces were paid $8.50 for each one, how much would they earn for their total production in one year?
   c) If it took approximately 5½ hours to create each timepiece or mechanism, how much time would it require to assemble all 100 million?
   d) If an average worker can produce 8 timepieces or timepiece mechanisms in a day, how many could he or she produce in a 235-day work year? How many workers would be needed to produce all 100 million?

LOCATION #177

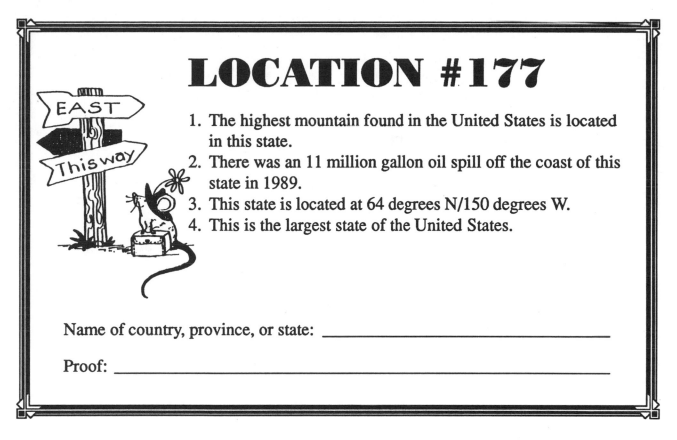

1. The highest mountain found in the United States is located in this state.
2. There was an 11 million gallon oil spill off the coast of this state in 1989.
3. This state is located at 64 degrees N/150 degrees W.
4. This is the largest state of the United States.

Name of country, province, or state: _____

Proof: _____

Creativity Across The Curriculum

In a team of three or four, create a game of Trivia Around The World. Do research to find important and interesting facts for at least 25 countries around the world. An example might be—"In what country would you find the Nile River?" (Egypt.) Or, "Where would you go to find an igloo?" (Northern Alaska.) Be sure to include questions about your own state, as well as questions that your classmates should know because of the countries and states that they have studied. When your team has completed its game, try it out on a group of friends.

LOCATION #178

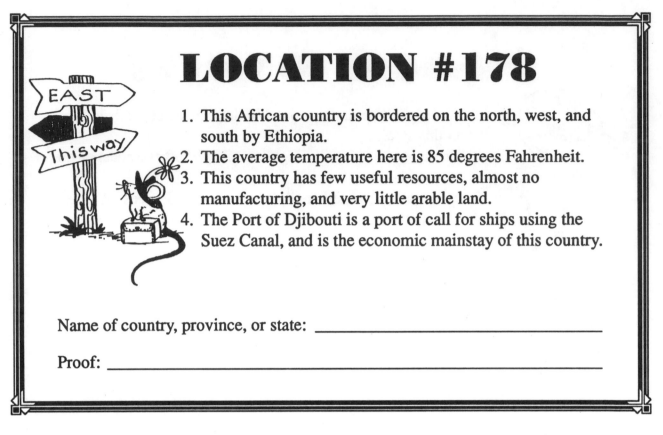

1. This African country is bordered on the north, west, and south by Ethiopia.
2. The average temperature here is 85 degrees Fahrenheit.
3. This country has few useful resources, almost no manufacturing, and very little arable land.
4. The Port of Djibouti is a port of call for ships using the Suez Canal, and is the economic mainstay of this country.

Name of country, province, or state: _____

Proof: _____

Creativity Across The Curriculum

1. List the five places you would most like to visit when traveling to Location Number 178. Give one interesting fact about each place you would visit.

2. Write a story about a fairytale character living in Location Number 178. How would the story change as a result of the different setting? For example, would Cinderella have married a tribal chief instead of Prince Charming? Sketch your fairytale character in this new setting.

3. What are the advantages/ disadvantages of living in an African city compared to a city in the United States?

Name _____

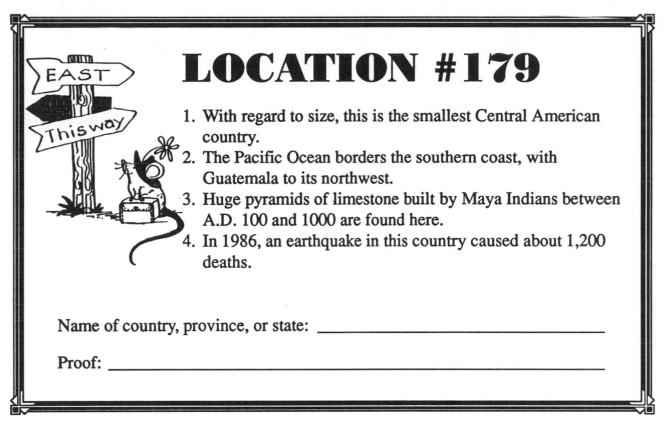

LOCATION #179

1. With regard to size, this is the smallest Central American country.
2. The Pacific Ocean borders the southern coast, with Guatemala to its northwest.
3. Huge pyramids of limestone built by Maya Indians between A.D. 100 and 1000 are found here.
4. In 1986, an earthquake in this country caused about 1,200 deaths.

Name of country, province, or state: _____

Proof: _____

Creativity Across The Curriculum

1. This country has often been involved in bitter conflict and war. What causes family conflict at home? How could you deal with this conflict in a positive way? Create a chart of conflicts and their possible resolutions.

2. Design a banner promoting peace among all people. Think of an original and provoking slogan for your banner.

3. Read *The Butter Battle Book* by Dr. Seuss. What does this book say to us about conflict?

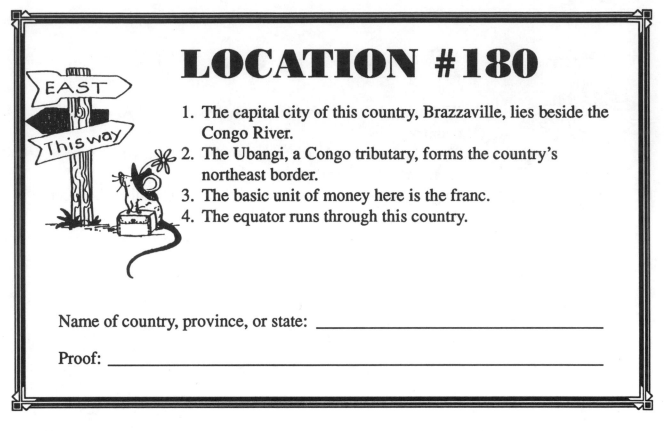

LOCATION #180

1. The capital city of this country, Brazzaville, lies beside the Congo River.
2. The Ubangi, a Congo tributary, forms the country's northeast border.
3. The basic unit of money here is the franc.
4. The equator runs through this country.

Name of country, province, or state: _____

Proof: _____

Creativity Across The Curriculum

1. It is the rainy season here and you have found a pot of gold at the end of the rainbow. In the pot of gold are 2,000,000 francs. How will you spend them?

2. You have to cross the Congo River at a very dangerous section. You do not have a boat. What will you do? (There is no bridge in the area.)

3. You are fishing in the Congo River with your dad's best fishing gear when you accidentally hook a crocodile. What will you do?

CREATIVE THINKING AND RESEARCH PAGES

Name _____

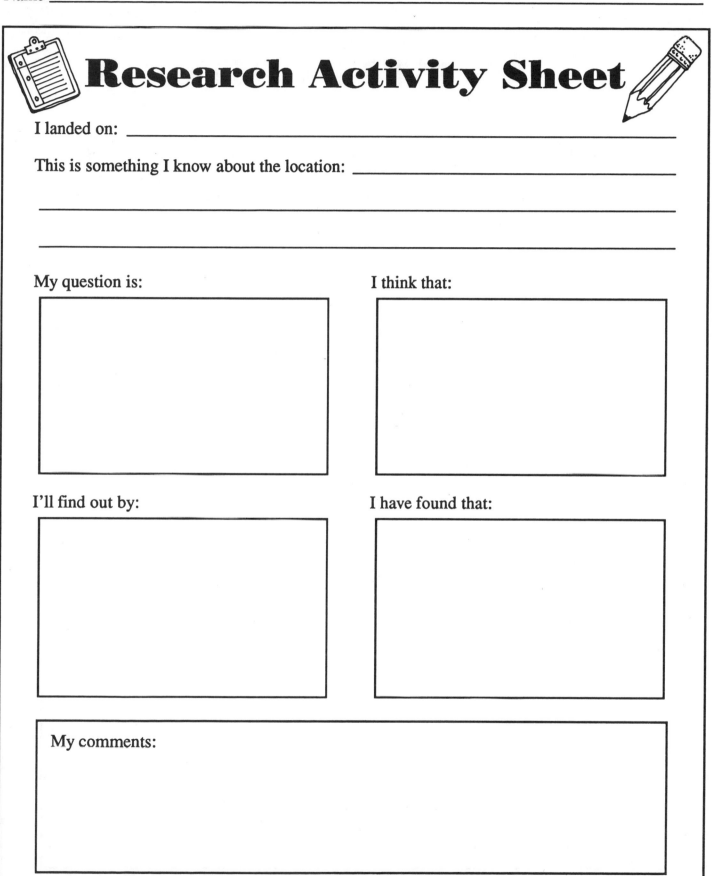

Research Activity Sheet

I landed on: _____

This is something I know about the location: _____

_____

_____

My question is:

I think that:

I'll find out by:

I have found that:

My comments:

ARTIFACT BOX

Prepare an artifact box that would help another student or students learn more about a particular country. You may use magazine pictures or real objects as artifacts to provide clues about the country's people, landforms, climate, economy, history, location, ancient and present-day customs, education, jobs, languages, and religions. **Place each information clue on a card. The information clue may be anything from a description of the location of the country using latitude and longitutde measurements to a list of the country's major products.** Put one or more artifacts and information cards together in a small plastic bag. See if a classmate can determine the country, based on your artifacts and information clues.

PYRAMIDS

INFORMATION CARD

Location Number _____

CLUE:

Name _____

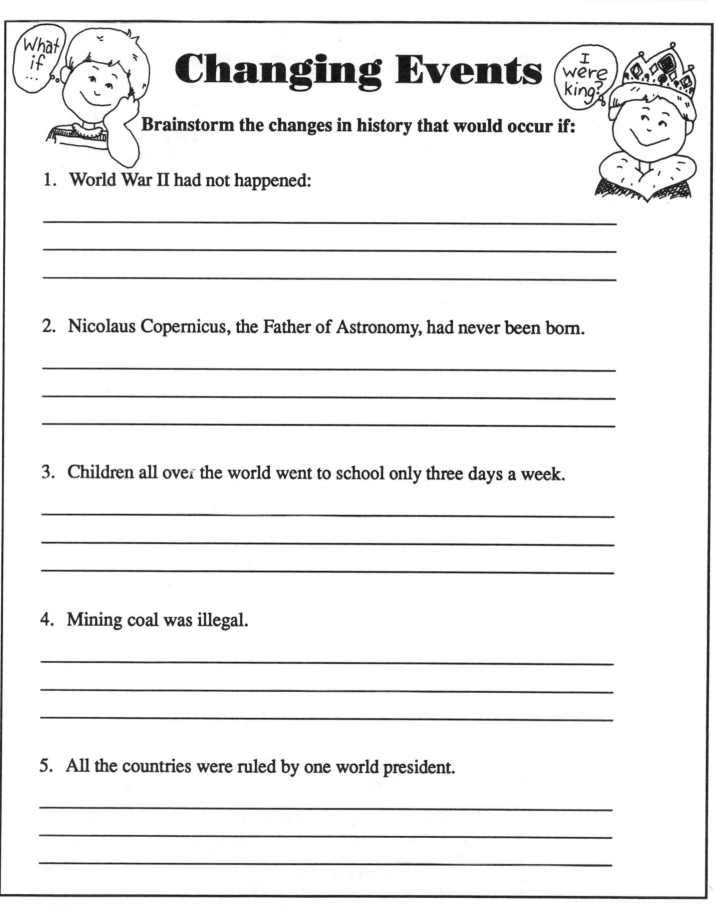

Changing Events

Brainstorm the changes in history that would occur if:

1. World War II had not happened:

_____

_____

_____

2. Nicolaus Copernicus, the Father of Astronomy, had never been born.

_____

_____

_____

3. Children all over the world went to school only three days a week.

_____

_____

_____

4. Mining coal was illegal.

_____

_____

_____

5. All the countries were ruled by one world president.

_____

_____

_____

What Time Is It?

You probably already know that the circumference of any circle is 360 degrees. Since the earth is a sphere, the distance around it at the equator is also 360 degrees. You also know that a day is made up of 24 hours, because the earth rotates on its axis, taking 24 hours to complete one rotation.

It follows, then, that the earth must rotate 15 degrees of longitude in each hour of time. (This is determined by dividing 360 degrees by 24 hours.) When you look at a globe you can see that time changes one hour for each 15 degrees. Because the earth rotates in a west-to-east direction, our day begins with the sun rising in the east and setting in the west. You can determine the time in any part of the world if you have a good map that will show you the number of longitudinal degrees between locations and the borders of time zones. If you are in Minneapolis, Minnesota, at 3:00 p.m., and telephone a friend in New York City, your friend might tell you that it is already 5:00 p.m. in New York.

Using this information regarding changes in time, see if you can complete the following problems.

1. How many hours' difference is there between Johannesburg, South Africa, and New York City, New York? _____

    If it is 12:00 noon in New York City, what time is it in Johannesburg? _____

2. How many hours' difference is there between Cape Town, South Africa, and Mexico City, Mexico? _____

    If it is 3:00 p.m. in Cape Town, what time is it in Mexico City? _____

3. If a plane leaves New York at 5:00 p.m. on Wednesday and flies to Johannesburg, South Africa, on a flight lasting ten hours, what time would the passengers wish to set their watches for upon their landing? _____

(continued on page 229)

Name _____

What Time Is It?

Page 2

4. Jody is a college student who is going to study in Europe. If she boards her plane in Dallas, Texas, at 11:00 a.m. on Monday and flies for 9 hours and 35 minutes, what time will it be in Paris when she lands? _____

5. Mandy is traveling from Cape Town, South Africa, to the United States. She boards her plane at 7:30 a.m. Three hours later, the plane stops in Dakar to pick up additional passengers and is on the ground for 1 hour. Eight hours later the plane again lands in New York City to refuel. The flight resumes after 30 minutes and lands in Denver 2 hours and 45 minutes later. What time is it in Denver when Mandy arrives? _____

6. If Alex flies from Moscow to Cairo on a flight that lasts for 2 hours and 50 minutes, when will he arrive in Cairo if he leaves Moscow at 9:00 a.m.? _____

7. What is the difference in time between Washington, D.C., and Berlin, Germany? _____

8. If a plane leaves Manila at 6:00 pm. and flies for 2 hours and 20 minutes, what time will it be in Bejing when the plane arrives? _____

9. Austin is traveling from Canberra, Australia, to Los Angeles, California. His plane leaves Canberra at 11:20 p.m. and arrives in Anchorage 5 hours and 25 minutes later for refueling. Flight resumes for 3 hours and 10 minutes before finally touching down in Los Angeles. What time will the clock in the L.A. airport report? _____

10. Natalie and Edward drove from Paris to Berlin in 8 hours and 25 minutes. If they left home at 6:30 a.m., what time is it in Berlin when they arrive? _____

11. In what direction would you travel if you went from Algiers to Cairo? _____ If it took 4 hours and 35 minutes to make the trip by plane, what time would it be in Cairo upon landing if you had left Algiers at 2:30 p.m.? _____

12. When the sun rises in Tokyo at 6:00 a.m., what time is it in New York City? _____

13. Dr. Jorg Bergham will fly from Beijing to Moscow on Friday for a conference. If the flight originates at 3:30 p.m. and takes 7 hours and 45 minutes, what time will it be in Moscow when he arrives? _____

Latitude, Longitude, and Distance

The distance around the center of the earth is approximately 25,000 miles. An imaginary circle drawn around the earth in an east-west direction is called the equator. This circle can be divided into 360 degrees. When dividing 25,000 by 360, we find that, at the equator, there are approximately 70 miles from one degree of longitude (an imaginary line of division running north and south) to another. Therefore, if one location is 30 degrees west longitude and another location is 20 degrees west longitude, the distance between the two can be calculated by multiplying 10 (the difference in degrees) times 70. You could estimate the distance between the two points to be about 700 miles. Remember, this is true only at the equator.

As the parallels of latitude shorten, the distance from one degree of longitude to another also changes. If you know the distance around a parallel of latitude, you can compute the distance between parallels of longitude in the same way as the 70-mile distance was determined. Compute the miles from one degree of longitude to another at each of the following parallels. (The number in parentheses is the approximate number of miles around the earth at each parallel.)

1. 10 degrees (24,800 mi.): _____
2. 20 degrees (23,400 mi.): _____
3. 30 degrees (21,600 mi.): _____
4. 40 degrees (19,080 mi.): _____

5. 50 degrees (16,200 mi.): _____
6. 60 degrees (12,600 mi.): _____
7. 70 degrees (8,640 mi.): _____
8. 80 degrees (4,300 mi.): _____

Once you have figured out the approximate distance from one degree of longitude to another at the same (or approximately the same) parallel, you can estimate the distance between cities that are on or near that parallel. For example, you will see on a globe that Adelaide, Australia, and Cape Town, South Africa, are both located about 33 degrees south latitude. If you know the number of miles per degree of longitude at the thirty-third parallel, you can calculate the approximate distance between the two. You will need to know the longitudinal location of each city and then determine the shortest route between the two. Your calculations may look something like this:

The distance around the 33rd parallel is approximately 20,850 miles. This can be divided by 360 to get the approximate distance from one degree of longitude to another. This is approximately 58 miles. We then determine that Cape Town is approximately 17 degrees west longitude and Adelaide is approximately 139 degrees west longitude. Traveling west, the distance between the two would be 122 degrees. If we multiply 122 by 58, we can determine the approximate distance between the two. See if you can figure out the answer.

Latitude, Longitude, and Distance

Page 2

Calculate the *approximate* distance between each of the following major cities:

Adelaide, Australia, and Santiago, Chile _____

Manila and Khartoum_____

Cairo and New Orleans _____

Toronto and Bucharest _____

New York City and Chicago_____

Lisbon and Tokyo_____

Bombay and Mexico City _____

Boston and Toronto _____

Pittsburgh and Lisbon_____

If you know that parallels are always the same distance
apart from one another, and that the distance between
parallels of latitude is about 70 miles, you can calculate
the distance between cities that are located at approximately the same longitude. Try your hand at estimating
the distance between each of the following cities:

Shanghai and Manila _____

Dublin and Lisbon _____

Port-au-Prince, Haiti, and Boston _____

Make up a few problems of your own for a friend to solve.

Mathematically Speaking

Understanding Currency Exchange Rates:

Money is considered to be any good or service that can be exchanged for another good or service. The name of this money and its value varies from country to country. In fact, the value of any country's money may vary from one day to another depending on a number of economic and political factors. Money helps to stabilize a country and also helps a person to plan and budget.

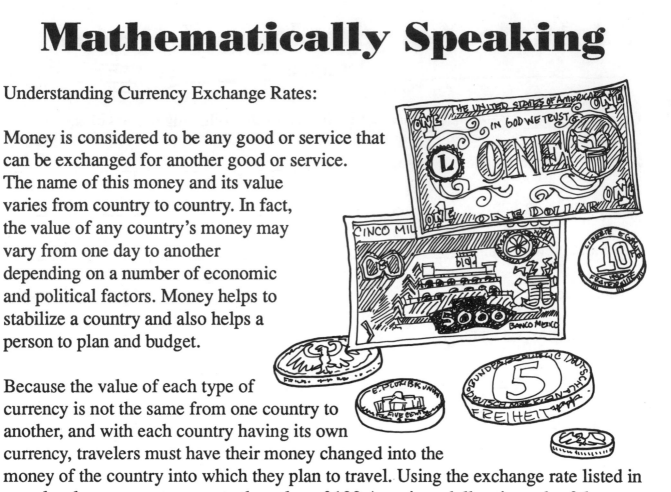

Because the value of each type of currency is not the same from one country to another, and with each country having its own currency, travelers must have their money changed into the money of the country into which they plan to travel. Using the exchange rate listed in your local newspaper, compute the value of 100 American dollars in each of the following currencies:

$100 = _____British Pounds

$100 = _____French Francs

$100 = _____Italian Lira

$100 = _____Japanese Yen

$100 = _____Greek Drachma

$100 = _____Mexican Pesos

$100 = _____Swedish Krona

$100 = _____South African Rand

$100 = _____Yugoslavian Dinar

$100 = _____Spanish Pesetas

$100 = _____German Marks

Flight Relay

Select a country. If you were flying to this country from your home state, what route would you take? Prepare the messages that the pilot would present to the passengers on the way. Be sure to talk about each point of interest that would be flown over.

FIRST RELAY

Example: "Good morning, ladies and gentlemen. It's nice to have you aboard flight #722, originating in _____ and destined for _____. Our arrival time is projected to be _____. As we pass over points of interest, I will inform you of their presence."

SECOND RELAY

"Ladies and gentlemen, we are now passing over_____,

flying at an altitude of _____ . You may notice

_____

_____

_____

_____."

THIRD RELAY

_____

_____

_____

_____."

Say What?

There are over 4,000 languages spoken in the world today. 1,000 of these are spoken on the continent of Africa alone. Do some research to unscramble the sampling of major African languages listed below. Try to find the country in which each language would be most frequently spoken.

	Language	Country
1. hlgneis		
2. aaasnfirk		
3. ulzu		
4. otohs		
5. muundb		
6. gnook		
7. carbia		
8. abarbma		
9. halsiiw		
10. yglaasam		
11. hnttoeoh		
12. silamo		
13. namhsub		
14. bererb		
15. kuama		
16. kuyuik		
17. sahox		
18. medne		
19. alaglin		
20. asiam		

Say What?
Page 2

Africa is the second-largest continent. It is divided into more than 50 countries. Among these 50 nations, over 1,000 languages are spoken. In what ways do you feel this might be a barrier to the development of Africa?

_____

_____

_____

_____

How might this further Africa's development? _____

_____

_____

_____

North America is the third-largest continent. How many languages can you think of that are spoken on the North American continent? Name them. _____

_____

_____

_____

How does this affect our country? _____

_____

_____

_____

If students in the United States were all required to learn a foreign language, what language do you think they should be required to learn, and why? _____

_____

_____

_____

What are some foreign words that you know? _____

_____

_____

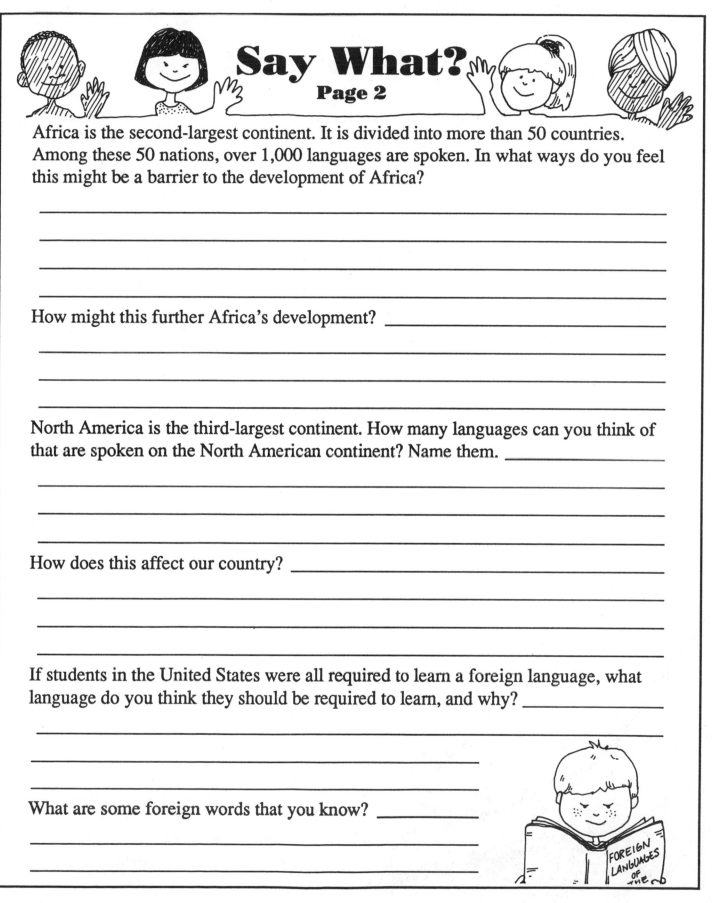

Eating Goober Peas

Although the Africans call them goobers and the Spaniards call them cacahuate, we simply call them peanuts in the United States. Mediterranean adventurers originally brought peanuts back from their travels to Peru and began to raise their own crops. Eventually Spanish and Portuguese explorers traded these crunchy treats with nearby Africa in exchange for spices and ivory.

The Africans enjoyed their "goobers" and even worshiped them along with other plants that they believed had souls. When these Africans were taken to America as slaves, their peanuts were put on board as a food source. Thus, peanuts arrived in America along with the slaves. The Americans, however, did not like them at first. They considered them to be food for the poor. It was not until the age of the circus and baseball that peanuts became popular with Americans.

George Washington Carver also played a large part in the development of the peanut's popularity. When cotton crops were destroyed by the beetle after the Civil War, farmers needed another crop to boost their income. Dr. Carver, a well-known black scientist, introduced them to the peanut. Thus, commercial peanut production began.

Do some research to find out how George Washington Carver further developed the use of the peanut. Create your own homemade peanut butter to reproduce the spread that revolutionized sandwich making in the U.S. All you need is a blender, peanut oil and peanuts.

PEANUT BUTTER

In the blender, combine 1 cup freshly roasted or salted peanuts with 1½ tablespoons peanut oil. Blend the mixture until it is smooth. Gradually add an additional 1½ tablespoons of oil. (Add ½ teaspoon salt if the peanuts are unsalted.)

Using your favorite jelly or jam, create your favorite peanut butter and jelly sandwich. How nutritious is a lunch that consists of a peanut butter and jelly sandwich, an apple, a glass of milk, and a cup of chocolate pudding? What is the most nutritious part of the lunch? What is the least nutritious part?

Significance Search

Read folktales from at least three different countries. Compare the basic elements of these folktales to the basic elements of American folktales. These include the explanation of natural occurrences, reflections of attitudes and ideals of the society, colorful characters, and expressions of moral value.

Answer Keys

Key for "Where In The World Am I?" (p. 37):
1. Harare (Salisbury)
2. Zimbabwe
3. Gabon
4. Kampala
5. Khartoum
6. Approximately 1800
7. Morocco
8. Rabat
9. Lisbon

Key for "Map Mania" (p. 130):
1. Northeast
2. Cape Town
3. North
4. Southeast
5. Lesotho
6. East
7. West
8. Pretoria
9. North
10. North
11. North
12. Northeast, South
13. Southwest, Northeast
14. Southeast

Key for Location 170 (p. 211):
Everest = 29,028 feet above sea level
McKinley = 20,320 feet above sea level
Difference between McKinley and Everest = 8,708 feet
Kilimanjaro = 19,340 feet above sea level
- Everest = 1,000 minutes McKinley = 700 minutes
- Difference = 9,688 feet

Key for "What Time Is It?" (p. 228):
1. 7 hours, 7:00 p.m.
2. 8 hours, 11:00 p.m.
3. 10:00 a.m.
4. 4:35 a.m.
5. 1:45 p.m.
6. 11:50 a.m.
7. 5 hours
8. 7:20 p.m.
9. 1:55 (p.m.)
10. 2:55 p.m.
11. east, 8:05 p.m.
12. 4:00 p.m.
13. 6:15 p.m.

Key for Latitude, Longitude, and Distance (p. 230; these are approximate answers):
1. 69 mi.
2. 65 mi.
3. 60 mi.
4. 53 mi.
5. 45 mi.
6. 35 mi.
7. 24 mi.
8. 12 mi.

Key for Latitude, Longitude, and Distance (p. 231; these are approximate answers):
1. 6030 mi.
2. 7200 mi.
3. 5145 mi.
4. 742 mi.
5. 8250 mi.
6. 11,180 mi.
7. 540 mi.
8. 3,975 mi.
9. 1050 mi.
10. 1050 mi.
11. 1750 mi.

Key for "Say What?" (p. 234—countries will vary, as many languages are spoken in several countries):
1. English — South Africa
2. Afrikaans — South Africa
3. Zulu — South Africa
4. Sotho — Botswana
5. Mbundu — Angola
6. Kongo — Congo
7. Arabic — Sudan, Egypt, Libya, Algeria, and others
8. Bambara — Mali
9. Swahili — Kenya, Uganda, Zaire, Tanzania, and others
10. Malagasy — Madagascar
11. Hottentot — Namibia (South West Africa)
12. Somali — Somalia
13. Bushman — Angola, Botswana
14. Berber — Morocco, Algeria, Western Sahara
15. Makua — Mozambique
16. Kikuyu — Kenya
17. Xhosa — Lesotho, South Africa
18. Mende — Liberia
19. Lingala — Central African Republic, Zaire
20. Masai — Kenya, Uganda

Alphabetical List Of Locations

Alaska 219
Albania 50
Alberta 102
Algeria 96
American Samoa 109
Angola 128
Argentina 150
Arizona 169
Armenia 28
Australia 193
Austria 199

Bahamas 39
Bahrain 206
Bangladesh 59
Barbados 116
Belgium 178
Benin 190
Bermuda 63
Bolivia 186
Bosnia 91
Botswana 74
Brazil 134
British Columbia 69
Brunei 155
Bulgaria 182
Burkina Faso 145
Burma 123

California 62
Cambodia (Kampuchea) 79
Cameroon 160
Canary Islands 45
Cape Verde 33
Central African Republic 139
Chad 23
Chile 57
China 80
Colombia 153
Colorado 90
Comoros 157
Congo 222
Connecticut 165
Costa Rica 110
Cuba 170
Cyprus 25

Delaware 187
Denmark 196
Djibouti 220
Dominican Republic 210

Ecuador 75
Egypt 146
El Salvador 221
England 214
Estonia 118
Ethiopia 179

Fiji 213
Finland 156
Florida 106
France 52

Gabon 136
Gambia 58
Georgia 200
Germany 125

Ghana 129
Greece 140
Greenland 103
Guatemala 64
Guinea 97
Guinea-Bissau 86
Guyana 70

Haiti 162
Hawaii 173
Honduras 175
Hong Kong 208
Hungary 183

Iceland 166
Idaho 205
Illinois 101
India 53
Indonesia 81
Iran 130
Iraq 188
Ireland 108
Israel 152
Italy 172
Ivory Coast 51

Jamaica 104
Japan 147
Jordan 137

Kenya 29
Korea 119
Kuwait 60

Laos 76
Latvia 40
Lebanon 46
Lesotho 73
Liberia 92
Libya 71
Lithuania 180
Louisiana 87
Luxembourg 158

Madagascar 98
Maine 68
Malawi 127
Malaysia 124
Mali 197
Malta 112
Manitoba 56
Massachusetts 122
Mauritania 141
Mauritius 138
Mexico 34
Minnesota 133
Monaco 201
Mongolia 24
Montana 65
Morocco 163
Mozambique 207

Nepal 211
Netherlands 176
Newfoundland 204
New Mexico 38
New York 114
New Zealand 184

Nicaragua 167
Niger 191
Nigeria 212
Norway 43
Nova Scotia 31

Oman 82
Ontario 174

Pakistan 113
Panama 151
Paraguay 189
Pennsylvania 100
Peru 67
Phillipines 105
Poland 217
Portugal 36
Puerto Rico 99

Quebec 85

Rhode Island 144
Romania 126
Rwanda 47

San Marino 94
Saudi Arabia 216
Scotland 26
Senegal 54
Sierra Leone 61
Somalia 88
South Africa 131
Spain 149
Sri Lanka 181
Sudan 120
Suriname 159
Swaziland 198
Sweden 142
Switzerland 218
Syria 66

Taiwan 168
Tanzania 203
Texas 83
Thailand 41
Tonga 192
Trinidad and Tobago 77
Tunisia 177
Turkey 72

Uganda 164
Utah 209
United States of America 185

Venezuela 48
Vietnam 95
Virginia 27

Wales 107
Washington D.C 32
Wisconsin 55
Western Sahara 78

Yugoslavia 89

Zaire 121
Zambia 84
Zimbabwe 42